Ethics in Science

Astha Saxena

Ethics in Science

Pedagogic Issues and Concerns

Springer

It is indeed fortunate, and of immense satisfaction, to see the serious yet accessible treatment that is afforded to SSI and related topics in Prof. Astha Saxena's new book. Although a good deal of time has passed (at least from my perspective) since my graduate days, the topic of ethics in science, while likely acknowledged as important to educators these days, is an area that when pressed for details, most have to confess ignorance. Ethics in science, most educators would assent, *seems* important, at least on prima facie grounds. But what exactly would constitute a reasonably working grounding in ethical inquiry, socioscientific issues or pedagogical content knowledge for this domain? What would the implementation of teaching strategies, scientific argumentation skills or actual classroom discourse look like? These questions are central to achieving functional scientific literacy. Thankfully, Prof. Saxena takes the reader on a journal of self-discovery in addressing all of these issues.

I find that most books written in related areas are not apprehensible to the practitioner either because of the density of verbiage or lack of direct applicability to their classroom needs, or in contrast, provide lock-step approaches to implementation without providing an appropriate rationale for the educational, philosophical or developmental importance for the exercise of such approaches. *Ethics in Science* strikes the Golden Mean. It does not shy away from philosophical topics such as Kant's Categorical Imperative, for example; rather, it is presented in a manner that helps provide a pedagogical rationale for its importance in science teaching and learning. Likewise, it provides a road map for teachers and students to understand the nuances of discourse through actual classroom narratives. I find this a refreshing contrast to the step-by-step (lock-step) approach because as with any road map, it allows for the exercise of judgement in deciding on how one might choose from a plethora of alternative routes to arrive at a certain destination in a journey that best suits their sensibilities.

This book is research-based but tempered by an intimate familiarity with practice. It extends the conversation about the central role of morality and ethics in science education. For the uninitiated, it provides a solid structure for the epistemic scaffolding of SSI and ethical discourse. For those who already have been working in this area, it offers a fresh and creative perspective to the power of teaching science-in-context that is personally meaningful. While this work focuses on secondary and post-secondary experiences, I believe that its underlying argument and principles are applicable to all levels of formal and informal science education. I invite the readers to begin their own journal no matter where the starting point lies.

Dana L. Zeidler
Distinguished University Professor
Science Education
University of South Florida
Tampa, USA

References

Bronowski, J. (1956). *Science and human values*. New York: Julian Messner, Inc.

Kahn, S., & Zeidler, D. L. (2019). A conceptual analysis of perspective taking: Positioning a tangled construct within science education and beyond. *Science & Education*. Early View: https://doi.org/10.1007/s11191-019-00044-2.

Zeidler, D. L. (2014). Socioscientific Issues as a Curriculum Emphasis: Theory, Research and Practice. In N. G. Lederman & S. K. Abell (Eds.), *Handbook of Research on Science Education, Vol. II* (pp. 697–726). New York, NY: Routledge.

Zeidler, D. L. & Keefer, M. (2003). The role of moral reasoning and the status of socioscientific issues in science education: Philosophical, psychological and pedagogical considerations. In D. L. Zeidler (Ed.), *The role of moral reasoning on socioscientific issues and discourse in science education*. The Netherlands: Kluwer Academic Press. (pp. 7–38).

Preface

Science and ethics can no more be viewed dialectically, as both have scope for the other. Science and its method can no more be taken as objective and empirical in approach; rather, they inhere an element of subjectivity coming mainly through researcher's bias, background, culture, prior knowledge, etc. However, ethics determines the code of conduct in the society and helps in decision-making as to which actions are deemed "good" along with the reason for choosing particular actions over others. The role of Ethics in Science comes via two different routes: One is through the content of Science where Ethics and ethical principles can be integrated for an interdisciplinary understanding of the issues that arise at the interface of Science, technology, society and environment, and the other way for integrating ethics in science could be indulging in the ethical practice of Science, wherein issues such as plagiarism, manipulation of data, honesty in recording of the observations, and issue of fair treatment and equal opportunity need to be addressed.[1] The former approach forms the base and background for the latter one, as the understanding of ethics, ethical principles and their applications is very important so that the practice of science becomes more ethical and less materialistic. Further, another component can be added to make the relationship complete:

Science-------Ethics--------Humanism

This component of Humanism or care for human individuals comes only when sciences abide by the ethical principles and clearly denotes that science without ethics cannot lead toward a humanitarian growth and development (Frolov & Yudin, 1989). These interlinkages between science, ethics and humanism form the basis of a socioscientific-ethical approach proposed for sciences. Many such incidents can be cited where the external agencies manipulated science to serve certain anthropocentric gains flouting all ethical codes and principles. Bhopal Gas Tragedy of 1984 was one of the most devastating incidents of the twentieth century in which

[1]Muralidhar, K., Ghosh, A. & Singhvi, A. K. (Eds.) (2019). Ethics in Science Education- Research and Governance. New Delhi:INSA.

the leakage of methyl isocyanide gas caused many deaths along with many unknown mutations resulting in serious abnormalities and defects in the future generations. The cause of that leakage, it is still a mystery, besides the Union Carbide Corporation (UCC) is still not willing to reveal the exact composition of the lethal gas cloud that was responsible for the carnage. This goes clearly against the scientific ethos and ethics of not communicating the results of the research carried out by UCC to the community. Similarly, another such incident, the Endosulfan Tragedy, occurred in Kerala in 2010 where the pesticide chemical caused many physical ailments in the people and the nearby farmers. Here, also, a sheer violation of research ethics is evident in not testing the chemical appropriately for its side effects on human population. All of these incidents rather tragedies resulted from a particular notion of development that gave more precedence to technological growth and human anthropocentrism. The prevalent notion needs to be checked by incorporating a synergistic interrelationship between developmental initiatives, society and environment. Hence, the moot question that should be of concern today is that, do we really need such technology? Is there an urgent need to take a detour so as to revive the lost human values and ethics that should have been the guiding force in each and every scientific and technology research?

Thus, a need arises at this time for the sciences to consider the inseparable role of ethics in its each and every endeavour, because it is only by virtue of ethics that sciences can be liberating in a true sense and transform the materialistic and anthropocentric mindset to an egalitarian and eco-centric one.

This component of 'ethics in science' was for the first time exhibited by Sir Robert K. Merton (1973), an American sociologist who founded the scientific sociology of science. He stated in his landmark paper on science and the social order, published in 1938, that science has an ethos that consists of intellectual honesty, integrity, epistemic communism, organized scepticism, disinterestedness, impersonality and universality (Richards, 1983). These represent a set of inherent value components of sciences and characteristic traits of a scientist that are considered to be the norms in a scientific community for carrying out authentic research practice. However, the purpose of the present study is not limited just to scientific research but to their application in the society that needs to be justifiable on ethical grounds. This is not in any way to undermine the kind of researches in science and technology but to generate an understanding about the real need, purpose and intent behind them that needs to be redefined and tested well so as to avoid such disasters to grip the humanity again.

This is what seems to be the missing link between science education and scientific literacy of which one component is an ethical understanding about day-to-day socioscientific issues. Thus, a need for including these issues in the curriculum at all the levels which need to be duly reflected in classroom teaching–learning processes is required to foster such an understanding. The present book is written with an aim to delve into the socioscientific-ethical character of science and technology and focuses upon the pedagogical aspect of ethical issues in the context of biological sciences.

The approach of the present book, *Ethics in Science: Pedagogic Issues & Concerns* is a bifocal one, as it aims to reinforce the role of Ethics in Science and also establishes the preponderance of teachers' knowledge domains in the teaching of ethical issues in biological sciences. An attempt has been made to highlight the place given to these ethical issues in the Science curriculum and praxis by understanding the pedagogy adopted and practised by the teachers in the Indian classrooms. This is the special feature of the present book. The theme of the book is situated broadly in the area of 'socioscientific-ethical issues', wherein an attempt has been made to delve into the deeper controversies and conundrums associated with various researches in science and technology and their impact on the society and environment. The curriculum at senior secondary and undergraduate levels though touches upon these areas by giving them some space in the textbooks, but the same might not be sufficient to generate an authentic ethical understanding on these issues. Hence, another purpose of this book is to highlight the place given to these issues in the textual content, classroom transactions, learner comprehension and assessment procedures.

The present book derives its significance from the fact that there are a very few reference materials available in the area of teachers' knowledge, notions, pedagogic practices and their interplay in the domain of socioscientific-ethical issues. This could be due to the fact that ethical issues are considered as vestigial part of the curriculum as it is a general notion that ethics and values tend to slow down the pace of development referred to as 'ethical lag' (Caplan, 1998). However, such a notion is now deeply contested as science is no more considered to be value-free and completely objective devoid of human component. Hence, the book in a way tries to reject this belief by explaining how values influence teachers' understanding of science and technology as well as taking decisions on the same when it impacts the society where ethics plays a predominant role. In the present century when we are at the pinnacle of technological progress and making a mark in the world be it any field such as medicine, agriculture, engineering, economics and nuclear power, then on the other hand we have also been a witness to some of the disasters of science and technology, such as Bhopal Gas Tragedy (1984), declining female sex ratio due to increasing feticides using amniocentesis[2], Endosulfan Tragedy in Kerala (1976–2000) and the recent Kerala Floods (2018). In the present book, an attempt has been made to understand the deeper contours of these tragedies and argue for the place of ethics in science and scientific practice. Can we continually use nature and its resources to the level beyond its carrying capacity? Do we need to be more mindful and ethical in our decisions with respect to introducing any new

[2]The worst hit region being Haryana, where such a practice is still prevailing under garbs. The statistics show that not only the illiterate but the educated elite are also going for such a practice due to a craving for a male child. Data from the census of 2001 suggests that there are only 933 women for every 1000 men in India (Census of India 2001 Provisional population totals http://www.censusindia.net). The UN in 2001 estimated that there were 44 million missing women in India (United Nations World population prospects: The 2000 revision: Highlights. New York: Population Division, Department of Economics and Social Affairs).

technology or for initiating developmental activities such as building of dams and urban settlement in a nature reserve or a sensitive area?

From the above examples, it also becomes clear that no technology can be placed in a void; rather, it is deeply embedded in the society which provides a platform for its applause or disdain. An authentic awareness in this regard can enable the citizens to make the 'right' choices and take the 'right' decisions without getting influenced by any propaganda. 'Ethics' is one area that can guide the process of decision-making and should constitute an integral component of science and technology. It has been felt that science education with all its intent and objectives must take this component into consideration which it broadly mentions as 'scientific literacy' in National Curriculum Framework (NCF 2005) developed by National Council of Educational Research and Training (NCERT), India, but fails to explain and elucidate. Thus, the present book in the backdrop of scientific literacy attempts to portray and emphasize the integration of science and ethics for a humanistic, philanthropic and responsible ideal of science education. This assumes even greater importance because of the search for international peace and harmony in contemporary world.

Astha Saxena, Consultant
Central Institute of Educational Technology
NCERT, New Delhi
Delhi, India

Acknowledgements

I express my deep gratitude to all those who have contributed in the publication of this book which has been a dream come true for me. I am indebted to Dr. Alka Behari, Professor, Department of Education, University of Delhi, New Delhi, India for providing me with useful insights for conceptualising the present book. I am grateful to Prof. Dana L. Zeidler, distinguished University Professor of Science Education, University of South Florida, Florida, USA for deliberating a foreword for the present book. I owe a sincere thanks to Prof. Namita Ranganathan, Head & Dean, Department of Education, University of Delhi, New Delhi, India for her patient reading of the drafts. The suggestions given by all of them have helped in bringing out the present version of the book. I am also thankful to Prof. Sadhna Saxena, Professor, Department of Education, University of Delhi, New Delhi, India and Prof. K. Muralidhar, Jawaharlal Nehru Chair, School of Life Sciences, University of Hyderabad, Hyderabad, India for their critical comments and valuable feedback on the book. Any errors and omissions are, of course, entirely my responsibility.

I owe a deep thanks to my peers and colleagues who have kindly read the chapters of the draft manuscript and offered constructive advice on how to improve the accuracy and clarity of the present book. In particular, my thanks go to Dr. Tulika Wadhwa and Dr. Manisha Subba. I am indebted to the teachers and students from different schools and higher education institutions in Delhi who had kindly consented to participate in the study and provided useful data for understanding the pedagogy of addressing ethical issues in the Science classrooms that has been delineated in the present book. I am grateful to my uncle, Dr. Anand Saxena for his valuable suggestions and motivation to write the present book. I also express my sincere gratitude to my aunts, Dr. Pravesh Saxena and Ms. Mrinal Kanta Saxena for their blessings and inspiration.

I am extremely thankful to Springer Nature and their team for publishing the current version of the book and providing me this great opportunity to present my work on an international platform and share my ideas with the world.

Finally, I would like to thank my parents who always encouraged me to write this book and supported me in every possible way.

New Delhi, India Astha Saxena

Contents

About the Author

Dr. Astha Saxena has done her Master's in Zoology and Education with a Ph.D. in the area of Science Education from Department of Education, University of Delhi, New Delhi, India. She has published papers, written articles, reports in the area of science education, teacher education and sociology and philosophy of science in reputed National and International Journals. Her key interest areas include Science Education, Science and Ethics, Bioethics, Environment Education, Teacher Education at elementary, secondary, senior secondary and higher education levels. She has been associated with organisations such as NCERT (National Council of Educational Research and Training), CBSE (Central Board of Secondary Education) and UNFPA (United Nations Population Fund) for developing Textbook on Pedagogy of Biological Sciences, MOOCs (Massive Open Online Courses) for Biology, Teacher Education and designing curriculum for Adolescence Health Programme. She was formerly teaching in Lady Irwin College, University of Delhi, New Delhi, India and is presently working as a Consultant at Central Institute of Educational Technology (CIET), National Council of Educational Research and Training (NCERT), New Delhi, India.

Abbreviations

AIDS	Acquired Immunodeficiency Syndrome
GM	Genetically Modified
GMO	Genetically Modified Organisms
KA	Knowledge of Assessment
KArg	Knowledge of Ethical Argumentation (Raising Arguments/Initiating Argumentation/Resolving Problems)
KCon	Knowledge of Context
KCur	Knowledge of the Curriculum
KET	Knowledge about Role of Ethics in Science
KNOS	Knowledge about Nature of Science
KSU	Knowledge of Students' Understanding
KTS	Knowledge of Transactional Strategies
NCERT	National Council of Educational Research and Training
NCF	National Curriculum Framework
OST	Orientation towards Science Teaching
PCK	Pedagogical Content Knowledge
SMK	Subject Matter Knowledge
SSI	Socioscientific Issues
STS	Science, Technology and Society
STSE	Science, Technology, Society and Environment
TB	Teacher's Beliefs

List of Figures

List of Tables

List of Boxes

Chapter 1
Science and Ethics—Merging the Boundaries

As a human activity, embedded in a context of Culture, and addressed to real human problems, science cannot possibly be value-free, or even ethics-free.
Bernard E. Rollin

Abstract Science and Scientific knowledge cannot be deemed value free as they have direct implications for the society where they are practiced and applied. Scientific research, including the applications of Science and Technology, often considered being a panacea for various ills of the society cannot be divorced from the culture and context where it breeds. The ethicists have often raised the question to whose benefit and by what means is Science and Technology progressing and would it really lead to the well-being and welfare of society at large? These are the questions often lurking whenever there is a clash of values, such as, in the case of Nuclear Power Plant, there is a clash of ethical values between establishing a country's military prowess and risking the lives of millions, or the case of GM crops, where there is a clash between self-sufficiency of a nation and human health. The present chapter addresses this conflict between Science and Ethics and attempts at building interrelationship between the two disciplines for a more conscious and responsible scientific research as well as for a harmonious living.

Keywords Science · Technology · Society · Ethics · Scientific research

1.1 Introduction

It is often said that science cannot be bereft of ethics. Even a simple experiment in science involving choice of scientific method in itself is a matter of value which the scientists often refuse to claim. Invasive use of animals in research for the purpose of human well-being and alleviation of human suffering is often touted as a moral claim, but is not valid in the case of human experimentation where human subjects are not used due to ethico-moral reasons. Similarly, funds are generally raised and provided for conducting research on AIDS (Acquired Immune Deficiency

A. Saxena, *Ethics in Science*, https://doi.org/10.1007/978-981-32-9009-9_1

Syndrome), Cancer, Pollution-free technologies rather than on say intelligence of frog, etc. which are based on socio-ethical decision-making (Rollin, 2006). Often, the scientists claim to be the proponents of pure knowledge ("myth of purity", rejected by Kitcher, 2001, p. 86 taken from Coulo, 2014) and abjure themselves from the applications (especially the disastrous ones) that their research could lead to. But researchers argue that scientists are to be held responsible for the consequences that their research could instil and should be responsible enough (Forge, 2008 cited in Coulo, 2014). This means that scientists cannot just keep on conducting research and producing new technology in the name of objective science but need to consider the repercussions it could have on the society and environment. Thus, the scientists need to be ethical enough to make the right decisions on their research and its applications as "ought" cannot be derived from "is" (which is a fact). This can be done by incorporating a set of non-cognitive values[1] in guiding their research (Coulo, 2014).

Thus, the notion that non-cognitive (non-epistemic) values form an integral part of sciences is being supported and argued for. An attempt has been made to critically analyse some of the recent biotechnological applications and ethical issues attached to each one of them. This is how interlinkages between science and ethics could be sought. Inclusion of this 'ethics' component has always been contested in Sciences which are no more considered as objective (Kuhn, 1970). But has it been able to achieve its due recognition or the axiom 'ethics in science' is still a farce? This calls for a realistic view of the whole situation, including the role of ethics in conducting experimentation in science labs, places where technology is used, created and distributed and most importantly in science classrooms where lessons and concepts related to such a technology are being taught.

1.2 Ethics: A Way of Life

Before we start our journey into the complex arena of ethical issues in sciences, it is important to get grips over the foundational level understanding about 'Ethics' per se. How do we define ethics? Is there any specific universal definition of ethics? Webster defines ethics as:

"The discipline dealing with what is good and bad and with moral duty and obligation 2a: a set of moral principles and values b: a theory or system of moral values c: the principles of conduct governing an individual or a group." The term

[1]Philosophical debates related to place of 'values' in science have generally bifurcated it into two different kinds, viz. cognitive (or epistemic) values that have become established within the scientific community and are practiced by the scientists. These epistemic (cognitive) values can be witnessed in reporting results of experiments, providing valid evidence, selecting problem, formulating hypothesis and choice of a valid theory to support the findings. The other kind of values which have not been duly regarded by scientists are a set of non-cognitive (non-epistemic) values. These include the moral and political values aiming at humanizing sciences (Coulo, 2014).

'Ethics' has originated from the ancient Greek word 'ethos' which means the moral purpose or character and is sometimes even used interchangeably with the word 'morals' concerned with differentiation of the goodness and badness of a particular human action (Chaffee, 2007). Ethics is often confused with emotions, reason or intellect, values and morals, culture and ethnicity, care and concern but is represented by none of them. Whenever a reference is being made using terms such as 'virtue', 'vice', 'duty', 'right', 'ought', 'good' or 'bad', then it pertains to ethical judgements, and if an analysis about their truth value is done then it is a point of discussion of ethics (Moore, 1993). Ethics in simple terms can be stated as the adherence to a particular mode of conduct that is considered permissible in a society at a particular time. This means that there can be a universal code of ethics that all of us should adhere to. To some extent yes, when it comes to the resolution of certain global issues but can be highly case sensitive and relative (ethical relativism[2]). Often it so happens, that conflict in opinion arises in any day-to-day situation or even at the level of global crisis where individual interests are at loggerheads, then it is here that methods of ethics and ethical mode of enquiry can only guide us towards the right decision-making.

1.3 Place of Ethics in Science and Science Education

Ethics is usually placed under the category of non-epistemic values that have a controversial place in science. This is because it does not directly contribute towards the pursuit of knowledge as also it is often in conflict with the more fundamental epistemic values of truth and objectivity (Gorham, 2011). The arguments that go against the integration of ethics and science are based on the fact–value distinction that was being upheld during the pre-Kuhnian era. This was based mainly on the processes and methods employed, which are very different for both the disciplines and can nowhere be reconciled. Ethical claims cannot be empirically verified as argued by logical empiricists. For example, Sciences have provided us with the means to generate nuclear power or say methods for crop improvement, but whether these should be applied or not is something which is outside the realm of sciences and enter into an ethico-moral domain. More so, according to some scientists, if sciences imbibe the ethical frames of references then it would lead towards a retrogressive or anti-technological development. On the other hand, another group of researchers argue that values and ethics are inseparable from sciences and have always been embedded in the processes of science, such as observance of truthfulness and honesty in reporting the results of an experiment, communicating one's findings amongst the fellow scientists (scientific communism), critiquing one's own findings (disinterestedness), universalism and

[2]'Ethical Relativism' is the claim that which actions are morally right varies from culture to culture and there is no such thing as objective rightness in morality (Rosenberg, 2012).

organized scepticism (Merton, 1973). Douglas (2000) also argues for the presence and relevance of both epistemic[3] and non-epistemic[4] values in science. Besides, the type of scientific researches that are being promoted is deeply affected by the ethos of the organization that supports and funds it. Various other extraneous factors co-determine the fate of researches that occur, such as vested interests of the stakeholders belonging and favouring a particular cadre of population, etc. (Reiss, 1999). An attempt to integrate this component of ethics in science via socio-scientific issues has been made by some science educators (Zeidler, Walker, Ackett, & Simmons, 2002; Goldfarb & Pritchard, 2000) so as to initiate ethical debates in the classrooms and foster an ethical understanding amongst the learners. With the discovery and creation of several new technologies, knowledge in the area of ethics becomes even more important, as the right decisions need to be taken regarding the ethically appropriate usage of a particular technology. Therefore, training in ethical ways of understanding, inquiry and decision-making is required in order to tackle these issues of global significance (Reiss, 1999).

Ethical issues are often considered as a vestigial part of the curriculum as it is a general notion that ethics and values tend to slow down the pace of development referred to as 'ethical lag' (Caplan, 1998). However, such a notion is now deeply contested as science is no more considered to be value free and completely objective devoid of human component. Hence, the present book in a way tries to reinforce this belief by studying how values and ethics influence our understanding of science and technology as well as taking decisions about the same when it impacts the society where ethics plays a predominant role. In the present century, when we are at the pinnacle of technological progress and making a mark in the world be it any field such as medicine, engineering, economics, etc., then on the other hand, we have also been a witness to some of the disasters of science and technology, such as Bhopal Gas Tragedy (1984), declining female sex ratio due to increasing feticides using Amniocentesis[5] in many parts of the country, Endosulphan Tragedy in Kerala (1976–2000) and the recent Kerala Floods (2018), drought in certain parts of the country and shifting patterns in seasons due to global warming are just a few to name.

[3]Epistemic values or the cognitive values refer to the properties of scientific theories such as simplicity, unification, accuracy, novel in prediction, explanatory breadth, empirical adequacy, internal consistency, connectivity, etc. (Doppelt, 2010) that help in building the knowledge base.

[4]Non-epistemic values refer to the social, ethical and political values that impact the scientific reasoning process (Douglas, 2000).

[5]The worst hit region being Haryana, where such a practice is still prevailing under garbs. The statistics show that not only the illiterate but the educated elite are also going for such a practice due to a craving for a male child (Data from the census of 2001 suggests that there are only 933 women for every 1000 men in India (Census of India 2001 Provisional population totals http://www.censusindia.net). The UN in 2001 estimated that there were 44 million missing women in India (United Nations World population prospects: The 2000 revision: Highlights. New York: Population Division, Department of Economics and Social Affairs).

From the above examples, it becomes clear that no technology can be placed in a void, rather it is deeply embedded in the society which provides a platform for its applause or disdain. An informed opinion in this regard would only enable the citizens to make the 'right' choices and take the 'right' decisions without getting influenced by any propaganda. 'Applied Ethics' is one area that can guide the process of decision-making and should constitute an integral component of science and technology. It has been felt that science education with all its intent and objectives must take this component into consideration which it broadly mentions as 'scientific literacy' as also given in our National Curriculum Framework (NCF, 2005) but fails to explain and elucidate. Thus, the present book in the backdrop of an interdisciplinary approach to teaching–learning Science and scientific literacy tries to portray and emphasize the integration of science and ethics for a humanistic and therapeutic ideal of science education. This assumes an even greater importance because of the search for international peace and harmony in contemporary world.

1.4 Ethical Frameworks for Research in Science and Technology

There are different ethical frameworks to rely upon and develop an ethical mode of thinking and decision-making with regard to ethical issues, such as Beauchamp and Childress (2008) have suggested four principles of ethical thinking: *Beneficence* (promoting good), *non-maleficence* (avoiding harm), *autonomy* (maximizing the freedom of individual and community) and *justice* (acting fairly). Similarly, Reiss (2003, 2006) presented four frameworks for ethical thinking, viz. consequences, autonomy, rights and duties, and virtue- or care-based ethics. No theory however in its own right is perfect and has many inherent problems that have been at times debated by ethicists and philosophers in the field. However, the aim of teaching ethics in science and technology should not be indoctrination of a particular viewpoint rather an informed moral reflection and debate where everyone's opinion is duly respected (Crosthwaite, 2001). The following ethical frameworks have been consulted for the present study based on which the data is interpreted and analysed.

1.4.1 Virtue Framework for Ethical Inquiry

The 'Virtue Ethical Theory' was for the first time given by Aristotle (384 B.C. to 322 B.C.) as a theory of moral conduct in the society. Which actions are considered good and moral and which one's bad or amoral formed the crux of the theory. But, the theory did not emphasize much on the consequences of particular actions, rather on the observance of a good and moral behaviour that can be acquired by choosing right action, by renouncing the desire for material objects and leading a moral and virtuous life. The theory of Aristotle firmly believes in 'Moderation' and not getting into the extremes of any behaviour or action (Kemerling, 2001) so that both too less and too much of an action are undesirable and a midway path should be adopted. Often the moral virtues are acquired by exercising them and making it a habit, such that a person is just if she/he performs just actions, brave if performs brave actions and the like (Aristotle, 1961).

1.4.2 Deontological Framework for Ethical Inquiry

The word 'deontology' is a word that is derived from two Greek words, 'deon' and 'logos' which mean duty and obligation, respectively (Freeman, 2002). The theory thus lays more stress on the observance of one's duty towards oneself as well as towards others, and the ends do not justify the means. One of the greatest proponents of this theory is Sir Immanuel Kant (1724–1804) as 'Kant's Categorical Imperative' of which two main principles are (a) Universalisability, such that an action which is taken up can be converted into a maxim that is universally applicable (Statement: "*Act only on that maxim whereby thou canst at the same time will that it should become a universal law*") and (b) the second principle is based upon the need to respect and treat humanity, both as in your person and the other, not as a means only but an end in themselves (Kay, 1997). Kant places goodwill or one's duty even above 'reason' or 'inclination' as otherwise a particular action borne from desire or inclination cannot be respected or aimed at. He argues that a true destination is to produce 'good will', such that an action derives its moral worth not from the purpose that it satisfies rather from the 'principle of volition' through which that action takes place (Kant, 1961). Kant lists four sources of moral motivation that are produced directly by reason, including, "moral feeling", "conscience", "love of human beings" and "respect" (Kant, 2002).

1.4.3 Utilitarian Framework for Ethical Inquiry

This particular framework gives more importance to the consequences of a particular action, which determine its morality. An act is considered as moral or ethical if it is supposed to benefit maximum number of people or at least cause no harm to others (Bentham, 1961). However, the method has been critiqued as who will determine the maximum number of people, is there a certain number? Does this number represent the share from different strata of society or only a specific group's interest/opinion? All these factors should be taken into consideration while applying this framework into day-to-day praxis. Besides, an alternative viewpoint propounded by one of the followers and disciple of Bentham, John Stuart Mill was to give greater credence to quality rather than quantity when accounting for happiness or pleasure from a particular action (Mill, 1961). In order to distinguish between greater good and higher good, an onus is being placed on the intelligence and a cultivation of nobleness of character, such that person with higher intellect and sound character does not derive happiness from transient pleasures, sensual indulgences to the injury of health, etc.

1.4.4 Pluralistic Framework for Ethical Inquiry

The ethical mode of enquiry should take due cognizance of a person's identity including her cultural values, tribe, gender, ethnic and religious perspectives for their share and role in ethical decision-making. Such pluralistic kind of societal structure leads towards pluralistic and divergent thought processes related to any particular issue, leading to disequilibrium and a state of unrest as every person suggests his/her own way of dealing with the problem and propose varied solutions. An informal dialogue and discussion can solve the matter by providing a platform for sharing of thoughts and opinions and reaching consensus without any indoctrination and power relationships. Science education without a pluralistic orientation cannot become liberating in a true sense as neglecting the opinion and practices of marginal groups will not only limit the source of our scientific knowledge but will tend to promote western science that is narrower in approach and intent as well as complete loss of our indigenous knowledge (Levinson, 2001; McGregor, 2004; Reiss, 1993).

1.5 Framework for Science Education

A growing awareness about the role of ethics in science for creating a scientifically literate citizenry has led many scientists as well as science educators to develop and devise such frameworks for teaching–learning that could help in fostering an ethical understanding amongst the learners. Earlier there was a wave of STS education introduced as a reform movement in science education pertaining to knowledge in the area of Science, Technology and Society (STS) issues that impinge our society and focussed upon the interlinkages of all the three components. Many science educators turned towards this approach (Aikenhead, 1993; Bybee, 1991) due to the simple reason that it helps in teaching and learning science beyond their scientific boundaries towards an interdisciplinary approach that encompasses the social world where the science and its practices are being applied and also helps in generating a critical viewpoint amongst the science learners. Later in the early twenty-first century, it was replaced by Socio-scientific Issues (SSI) that subsumes the earlier concept of STS and involves an in-depth analysis of the issues from a social, political, scientific and ethical perspective. Many teaching models have been proposed to tackle these SSI in the classrooms but it is usually felt that they do not receive their due recognition in the curriculum. It was also found that teachers usually hesitate from touching upon these issues in the classroom as they are highly controversial and difficult teaching areas (Levinson and Turner, 2001).

In order to understand the inherent complexity of SSI, the present textbook focuses upon some such socio-scientific issues within the curriculum of Biological Sciences across senior secondary and undergraduate level. An attempt will be made to take up each issue separately, discuss the sociopolitical, economic and ethical arguments and debates centred on them. These arguments and debates will then be situated within the broader ethical framework so as to carry out an ethical analysis and assessment therein. As there could be no one unanimously accepted viewpoint, thus each argument will be weighed differently by different subjective individual assessors (Mepham, 2008).

1.6 Bioethics—A New Branch of Study Linking Biosciences and Ethics

Bioethics is the study of ethical issues emanating from advances in the field of Biology and Medicine. The word "Bioethics" was proposed in 1970 by a Biological scientist, Van Rensselaer Potter to name his vision of a new connection of scientific knowledge with moral considerations of the converging evolutionary understanding of humans in nature (Jonsen, 2010). This was at a time during the late twentieth century when some thoughtful scientists became increasingly concerned about the social implications of their scientific research, especially those working in the field of health care, Genetics and Atomic energy. This led to the establishment of the Kennedy Institute of Ethics (1971) at Georgetown University pioneering a new area of research into medical ethics which was later preferred to be called by its founder, Andre Hellegers, as 'Bioethics'. In 1974, Dan Callahan, founded another major institute of research called 'The Hastings Centre' suggested the emergence and development of 'Bioethics' as a new Disciplinary area, wherein both the traditional methods of philosophical analysis and debate are combined with sensitivity, human emotion and social and political influences with which Medicine was practiced. The field of bioethics addresses a broad range of issues and debates over the boundaries of life (e.g. abortion, euthanasia), In Vitro Fertilization (IVF), the allocation of scarce healthcare resources (e.g. organ donation, healthcare rationing) to the right to refuse medical care for religious or cultural reasons. With increase in the scope and intrusion of technology into human life covering Biotechnology which includes cloning, genetic engineering, gene editing, organ regeneration and transplantation and life in space, the need for an inter-discipline such as Bioethics becomes indispensable. It is here that the people will need to make the 'right' decisions that do not hamper their own life as well as the lives of those around them. This not only requires an understanding of the scientific aspects of a particular technology rather the sociological, political, economic, emotional and ethical aspects.

The three main tasks of a bioethicist include definition of issues, methodological strategies and procedures for decision-making. Each of these tasks is rigorous and requires understanding of the basic principles (Callahan, 2010). Wherein, the first one requires deeper insights into the issues, ability to see in and through many different perspectives. The second one involving the method requires multiple approaches such as philosophical, teleological, scientific, psychological and emotional. The decision-making task of a bioethicist is directly linked to the methodology; in fact, it is the nature of method employed while resolving a particular ethical problem that gives an idea as to how can one reach reasonably specific and clear decisions. It is not only the bioethicist who needs to take a decision with regard to application of certain technologies but also important on the part of a teacher who is involved in the teaching–learning processes associated with these technologies. As today's students would be tomorrow's scientists, engineers and medical practitioners; therefore, it becomes all the more imperative for them to think critically about these technologies so as to reach the 'right' decisions.

The aim of the present book is not to provide some solutions or ready-made answers to the complex ethical issues associated with various technologies rather it is more about cajoling the reader into the process of critical and analytical thinking on these issues and provide them alternative paths to consider and reflect. The following cases (Boxs 1.1, 1.2, 1.3 and 1.4) are evidence for the unbridled growth of Science and Technology leading to some of the worst disasters that mankind has faced. Such instances compel us to think and reflect on the route to development that the countries are adopting to reach the pinnacle severing the human–environment ties and responsibilities.

Box 1.1 Bhopal gas tragedy and its aftermaths

Bhopal gas disaster girl, the burial of one iconic victim of the gas leak (4 December 1984) (*Source* https://en.wikipedia.org/wiki/Bhopal_disaster#/media/File:Bhopal_gas_disaster_girl_by_pablo_bartholomew.jpg)

The Bhopal Gas Tragedy is said to be the major gas leakage disaster that happened on the intervening night of 2–3 December 1984 at the Union Carbide India Limited (UCIL) pesticide plant in Bhopal, Madhya Pradesh, India. It is considered to be the World's worst industrial disaster where around 5,00,000 people were exposed to Methyl Isocyanate (MIS) gas. Neither the factory owners nor the Government had any idea to respond to such a deadly gas. The Union Carbide Corporation (UCC) that owned the plant also did little to deal with the human tragedy. It is being said that the Union Carbide used trade secrecy as their prerogative and withheld information on the exact composition of the leaked gases (Narain & Bhushan, 2015). Even till late 2014, ICMR (Indian Council of Medical Research) could not detect the real cause of tragedy and the diseases that followed after the disaster. The disaster that happened affected the people and the environment in multiple ways, first, the victims who were directly exposed to the deadly gas, then the foetuses in the womb who were indirectly affected by the gas, and also the chemical waste that was dumped near the factory contaminating the water that the people drank. The disaster that could have been averted through proper planning and management raises many ethical issues such as

Points of Discussion:

- First, what was the need to set up a pesticide plant in an area cohabited by a thick population?
- Could the leak be prevented and checked?
- Why at all did UCC select Indian state to put up the pesticide plant?
- Was it ethical to set up an industry while knowing about the dangerous chemicals used and released therein?
- Can the legal battle and settlement grants replace the agony and massive loss incurred on the people of Bhopal?

Box 1.2 Endosulphan tragedy in Kerala

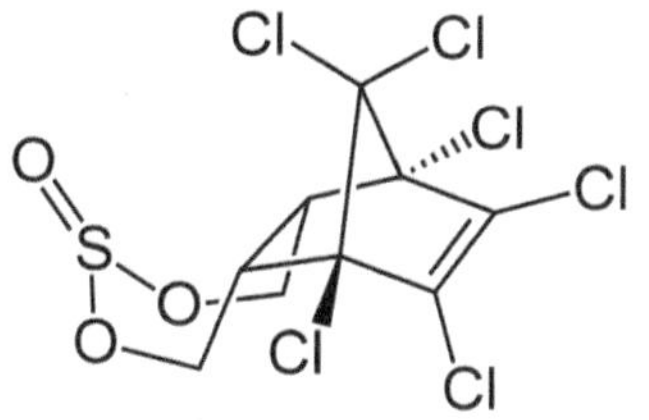

Endosulphan is an off-patent Organochlorine insecticide and acaricide that was sprayed indiscriminately over the fields to accelerate Cashew plantation in the state of Kerala from 1978 onward to increase the crop yield. This was especially taken up by the then Plantation Corporation of Kerala Ltd. and its mandate of spearheading the agro-economic growth in the state. Due to the mass aerial spraying of the chemical, its gruesome effects started showing as early as 1979, resulting in stunted growth, and deformed limbs, congenital anomalies, mental retardation, physical deformities, cerebral palsy, epilepsy, hydrocephalus foetuses among newly born and children. Soon, the Endosulphan Spray Protest Action committee was established as the people along with some doctors were able to figure out the reason for the maladies faced by the people of nearby villages where Endosulphan was sprayed. The residents in the area are the ones who have witnessed the trail of ecological destruction since the spraying of the deadly Endosulphan. One of the older resident in the area narrated his agony in the following manner,

"I don't see jackals in this area now," says Kajampady Subramanya Bhat, 75, whose family has lived in Padre since he doesn't know when. "In 1962-63, when the plantation started, they used to put groundnut cakes in the pits. A worker told me they were mixing pesticides in the cakes to prevent jackals from eating them. I don't see any jackals now, nor too many frogs, fish or crows. Rat snake, a farmer's friend as it checks the rodent population, has disappeared. There are no fireflies. I used to have 22 beehives. Now I have none." (Mishra & Joshi, 2018).

The Endosulphan tragedy again raises a number of ethical questions associated with the scientific research and its application.

- Why was the chemical (Endosulphan) sprayed indiscriminately over a thickly populated and waterbody-rich area like Kasaragod without estimating its prospective harmful effects on other organisms (including humans) and environment?
- Do the scientists associated with the discovery of this chemical (Endosulphan) be held responsible for the tragedy that occurred in Kerala?
- How could the Govt. officials allow such a rampant use of the Chemical without even knowing about its harmful effects?
- Could this tragedy be averted by any means?

Box 1.3 Deluge in Kedarnath, Uttarakhand, India

Rear view of the Kedarnath temple in the aftermath of the flood (*Source* https://en.wikipedia.org/wiki/Kedarnath_Temple#/media/File:Temple_in_the_aftermath_of_the_flood.jpg)

The Kedarnath Valley along with other parts of Uttarakhand was hit severely by the flash floods on 16 and 17 June 2013. An aftermath of which, thousands of pilgrims and locals in the valley were stranded leading to a death toll of around 5700 people. The reason for the flash floods as stated by the MET department was the unusually high rate of rainfall in the preceding days which led to cloud bursting and landslides. However, the environmentalists carried out research and concluded that one of the reasons for the disaster was actually man-made, owing to expansion of hydel projects, roads and tourism that is destabilizing the natural ecosystem of the state and hampering its sustainability (Basu & Singh, 2015). The following ethical concerns emerge from the above case:

- Should the construction projects be sanctioned at a sensitive place like Kedarnath which is prone to floods?
- In between 'Development' and 'Environment Conservation' which one is more important and taken care of?
- Could the disaster be averted by any means?
- Why was not the carrying capacity of the Himalayas assessed before planning huge projects?

Box 1.4 Designer babies in China

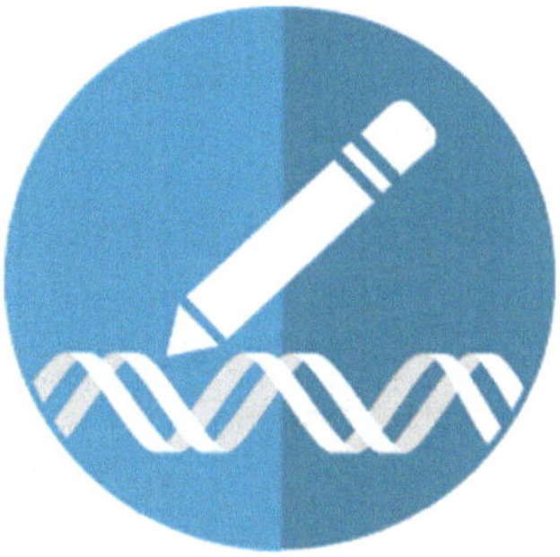

(*Source* https://pixabay.com/en/gene-editing-icon-crispr-icon-2375787/)

A Scientist in China claims to have created the World's first genetically edited babies. The technology 'gene editing' permits the selection, insertion or deletion of certain genes to repair any defective gene which then comes under the purview of 'gene therapy' but in the case of 'designer babies' a deliberate attempt is being made to select favourable genes and insert them in the genetic material of the embryos. Such a practice is legally banned and is even condemned by the scientists as it is still at an experimental stage and is unconscionable. It raises many serious ethical concerns such as

- Is it ethical to genetically modify or alter the genome of a human embryo just to select a better trait?
- What happens if the experiment fails?
- What if the designer baby develops some altered and unforeseen effects of gene editing?
- Does one human have any right to manipulate the genome of the other without taking consent?

From the above cases, it becomes clear that having a scientific knowledge base may not be sufficient to reach the 'right' decision with respect to a particular technology or its use. Rather, it is only when Science integrates Ethics or the ethical principles within its disciplinary domain that would lead to a philanthropic, responsible and therapeutic ideal of Science. Thus, Ethics becomes an indispensable element of science and its application. Thus, there could be many issues at the interface of Science, technology, society and environment where the chief aim is either development or enhancement of luxury or comfort. But what happens in the long run, or even near future, that nobody can predict as nature is a complex system and has its own way of functioning and reacting to any change. This has been beautifully put by E. F. Schumacher as follows:

> Scientific or technological 'solutions' which poison the environment or degrade the social structure and man himself are of no benefit, no matter how brilliantly conceived or how great their superficial attraction. Ever bigger machines, entailing ever bigger concentrations

> of economic power and exerting ever greater violence against the environment, do not represent progress: they are a denial of wisdom. Wisdom demands a new orientation of science and technology towards the organic, the gentle, the non-violent, the elegant and beautiful. Peace, as has often been said, is indivisible—how then could peace be built on a foundation of reckless science and violent technology? We must look for a revolution in technology to give us inventions and machines which reverse the destructive trends now threatening us all. (Schumacher, 1973)

1.7 The Structure of the Book

The present book is divided into three parts that have a logical sequence and deliberately arranged in an order. Each part of the book contains some chapters. The book opens with *Part-I, Ethics in Science Curriculum* that discusses the role of Ethics in Science and attempts to address the issue of integration of Ethics within the Science curriculum. It contains a chapter on Ethics in Science—an Inquiry into Bioethical Issues (2) which aims at developing interlinkages between science and ethics through the lens of various Bioethical issues such as Amniocentesis, Assisted Reproductive Technologies (ARTs), cloning, etc. The next chapter is about 'Place of Ethical Issues in the Science Curriculum—A case of K-12 Biology Textbook' (3) which presents a critical content analysis of the K-12 Biology textbook with respect to the inclusion of Bioethical issues thereby highlighting such areas in the Biology Text where there is some confluence of science, technology and Ethics as well as to critically analyse the content for generation of a critical thinking and ethical understanding on the same. The next chapter in this part is 'Pedagogical Approaches to Ethical Issues' (4) that presents a review of related literature in the area of pedagogy for addressing ethical issues in a Science classroom. It also introduces the concept of 'Pedagogical Content Knowledge (PCK)' and how the same is helpful in understanding a relationship between pedagogy and ethical issues for their effective transaction in the classroom. The second part of the book presents the *Case Narratives from India* (5–9) that includes case studies of five teachers who are teaching Biology at K-12 (two case studies) and Undergrad level (three case studies). Each teacher brings in some unique perspective to address these ethical issues in their respective classrooms. The detailed case studies are presented in a narrative style including some excerpts from teachers' own voices and their classroom discourses. *The third part of the present book presents the Tools for Learning which* includes two chapters, one on Ethical Understanding among students about Bioethical Issues (10) and another on Assessment for Ethical Understanding—Tools, Techniques and Strategies (11). Both these chapters help in gauging students' conceptions and assessing their understanding about ethical issues related to Biotechnology.

References

Aikenhead, G. (1993). Foreword: Multicultural issues and perspectives on science education. *Science Education, 77*(6), 659–666.

Aristotle. (1961). Moral and intellectual virtues. In M. K. Munitz (Ed.), *A modern introduction to ethics: Readings from classical and contemporary sources* (pp. 131–167). USA: The Free Press, a Corporation.

Basu, S., & Singh, J. (2015, July 8). Man-made reasons for Uttarakhand disaster. *Down to Earth*. Retrieved from https://www.downtoearth.org.in/news/natural-disasters/man-made-reasons-for-uttarakhand-disaster-41407.

Beauchamp, T., & Childress, J. (2008). *Principles of biomedical ethics* (6th ed.). New York: Oxford.

Bentham, J. (1961). The Principle of Utility. In M. K. Munitz (Ed.), *A modern introduction to ethics: Readings from classical and contemporary sources* (pp. 185–190). USA: The Free Press, a Corporation.

Bybee, R. W. (1991). Planet earth in crisis: How should science educators respond? *The American Biology Teacher, 53*(3), 146–153.

Callahan, D. (2010). The Emergence of Bioethics as Discipline and Discourse. In N. Jecker, A. R. Jonsen, & R. A. Pearlman (Eds.), *Bioethics: An introduction to the history, methods and practice* (2nd ed., pp. 17–22). Canada: Jones and Bartlett Publishers Inc.

Caplan, A. (1998). *Am I my brother's keeper?* (p. 256). Bloomington: Wiley.

Chaffee, J. (2007). *Thinking critically* (9th ed.). New York: Houghton Mifflin Company.

Coulo, A. C. (2014). Philosophical dimensions of social and ethical issues in school science education: Values in science and in science classrooms. In M. Matthews (Ed.), *International handbook of research in history, philosophy and science teaching*. Dordrecht: Springer Science + Business Media. https://doi.org/10.1007/978-94-007-7654-8_33.

Crosthwaite, J. (2001). Teaching ethics and technology: What is required? *Science & Education, 10,* 97–105.

Doppelt, G. (2010). Values in Science. In S. Psillos & M. Curd (Eds.), *The Routledge companion to philosophy of science* (pp. 302–313). New York: Routledge.

Douglas, H. (2000). Inductive Risk and Values in Science. *Philosophy of Science, 67*(4), 559–579.

Forge, J. (2008). *The responsible scientist*. Pittsburgh, PA: University of Pittsburgh Press.

Freeman, S. (2002). Deontology. Retrieved from http://www.routledge-ny.com/ethics/entries/deontology.pdf.

Goldfarb, T., & Pritchard, M. (2000). *Ethics in the science classroom: An instructional guide for secondary school science teachers with model lessons for classroom use*. Retrieved from www.wmich.edu/ethics/ESC/index.html.

Gorham, G. (2011). *Philosophy of science: A beginner's guide*. England: Oneworld Publics.

Jonsen, A. R. (2010). A history of bioethics as a discipline and discourse. In N. Jecker, A. R. Jonsen, & R. A. Pearlman (Eds.), *Bioethics: An introduction to the history, methods and practice* (2nd ed., pp. 3–16). Canada: Jones and Bartlett Publishers.

Kant, I. (1961). The categorical imperative. In M. K. Munitz (Ed.), *A modern introduction to ethics: Readings from classical and contemporary sources* (pp. 258–272). USA: The Free Press of Glencoe.

Kant, I. (2002). *Groundwork for the metaphysics of morals*. (A. W. Wood, Ed.). New York: Vail-Ballou Press.

Kay, C. D. (1997). Kantian ethics. Retrieved from www.wofford.edy/kaycd/ethics/deon.htm.

Kemerling, G. (2001). Philosophical ethics: Aristotle. Retrieved from www.philosophypages.com/ph/aris.htm.

Kitcher, P. (2001). *Science, truth and democracy*. Oxford: Oxford University Press.

Kuhn, T. S. (1970). *The structure of scientific revolutions* (2nd ed.). USA: University of Chicago Press.

Levinson, R. (2001). Should controversial issues in science be taught through the humanities? *School Science Review, 82*(300), 97–102.
Levinson, R., & Turner, S. (2001). *Valuable lessons: Engaging with the social context of science in schools*. London: The Wellcome Trust.
McGregor, D. (2004). Coming full circle: Indigenous knowledge, environment and our future. *American Indian Quarterly, 28,* 385–409.
Mepham, B. (2008). *Bioethics: An introduction for the biosciences* (2nd ed.). New York: Oxford University Press Inc.
Merton, R. (1973). *The sociology of science: Theoretical and empirical investigations* (pp. 267–278). Chicago: The University of Chicago Press.
Mill, J. S. (1961). What utilitarianism is? In M. K. Munitz (Ed.), *A modern introduction to ethics: Readings from classical and contemporary sources* (pp. 191–205). USA: The Free Press, a Corporation.
Mishra, S., & Joshi, S. (2018, August 16). Tracking decades-long endosulfan tragedy in Kerala. *Down to Earth.* Retrieved from https://www.downtoearth.org.in/coverage/health/tracking-decades-long-endosulfan-tragedy-in-kerala-56788.
Moore, G. E. (1993). *Principia Ethica* (2nd ed.). U.K: Cambridge University Press.
Narain, S., & Bhushan, C. (2015, July 08). 30 years of Bhopal gas tragedy: A continuing disaster. *Down To Earth.* Retrieved from https://www.downtoearth.org.in/coverage/environment/30-years-of-bhopal-gas-tragedy-a-continuing-disaster-47634.
Position Paper National Focus Group on "Teaching of Science." (2005). New Delhi: NCERT.
Reiss, M. (1993). *Science education for a pluralist society.* Bristol: Open University Press.
Reiss, M. J. (1999). Teaching ethics in science. *Studies in Science Education, 34*(1), 115–140.
Reiss, M. (2003). How we reach ethical conclusions. In R. Levinson & M. Reiss (Eds.), *Key issues in bioethics* (pp. 14–23).
Reiss, M. (2006). Teacher education and the new biology. *Teaching Education, 17,* 121–131.
Rollin, B. E. (2006). *Science and ethics.* New York: Cambridge University Press.
Rosenberg, A. (2012). *Philosophy of science: A contemporary introduction* (3rd ed.). Taylor & Francis.
Schumacher, E. F. (1973). *Small is beautiful: A study of economics as if people mattered.* London: Blond & Briggs Ltd.
Zeidler, D., Walker, K., Ackett, W., & Simmons, M. (2002). Tangled up in views: Beliefs in the nature of science and responses to socioscientific dilemmas. *Science Education, 86,* 343–367.

Websites

http://endosulfanstudy.blogspot.com/2012/10/v-behaviorurldefaultvmlo.html.
http://www.indiaenvironmentportal.org.in/files/IndiaEndosulfan.pdf.
https://www.theguardian.com/science/2018/nov/26/worlds-first-gene-edited-babies-created-in-china-claims-scientist.

Part I
Ethics in Science Curriculum

Chapter 2
Ethics in Science: An Inquiry into Bioethical Issues

Abstract Science and ethics can no more be viewed dialectically, which has been deliberated by many scientists and philosophers who have contributed to the socio-scientific understanding of various scientific inventions and researches. The emergence of a new branch of study, 'Bioethics' that deals with the ethical questions that arise at the interface of Biosciences, medicine, various Biotechnological applications, politics and law is an evidence for the growing interrelationship between the twin disciplines for the welfare and well-being of the society and environment. The present chapter delves into some such Bioethical issues, such as animal experimentation, amniocentesis, In Vitro Fertilization (IVF), genetic engineering, etc. The basic idea is to explore these issues from an ethical perspective to reveal the unspoken or hidden equation between Science and Ethics, and also to determine how decision-making in Science and scientific research could be disastrous if not done in an ethical manner.

Keywords Science · Ethics · Bioethics · Well-being · Bioethical issues

2.1 Introduction

The present chapter delves into one of the emerging and interdisciplinary areas of science education which has never been given its due recognition and place within the Science curriculum and praxis. There is no specific terminology to define and delimit this virgin area of science education which has immense scope for inquiry and deliberation. This is the area of Science and Ethics, an alliance between two entirely different and rather converse disciplines both in their processes and intent. Although subjectivity in science has been approved according to many philosophers of science (Kuhn, 1970; Chalmers, 1999), it still does not embrace the idea of 'ethics' which provides us with the knowledge about 'right' and 'wrong' along with the reasons behind choosing particular actions over others. However, this dichotomy between science and ethics has been deeply contested, and the conventional view about value-free science has been rejected. A common forum where both the disciplines merge together is the realm of Socio-Scientific Issue (SSI) that provides

A. Saxena, *Ethics in Science*, https://doi.org/10.1007/978-981-32-9009-9_2

a forum for discussion and debates on some of the issues related to science and technology that impact our society. The present chapter revolves around some such ethical issues that have emerged in recent times owing to the sudden spurt in certain science and technology researches leading to an unprecedented impact on organismic life and environment in general. These have been referred to as ethical issues in the present chapter rather than socio-scientific issues as the stress is more on the ethical aspect inherent in them.

2.2 Ethical Issue Analysis—The Method and Approach

In order to analyse the issues from an ethical perspective, reference to three major ethical frameworks viz., virtue framework, deontological framework and utilitarian framework can be drawn. This was made possible by making use of an 'ethical matrix' (Mepham, 2008), wherein the contributions of different stakeholders could be made very evident and explicit at one place. Such a procedure aids in right kind of decision-making and saves from getting biased towards any particular group of people. Such a matrix allows the presentation of diverse arguments and viewpoints as well as identifies any clash of values, ethical stances and opinions among different stakeholders. The ethical issue analysis once undertaken and understood will foster a better understanding and analysis of the Biology textbook of K-12 for the inclusion of ethical issues. It will also provide a backdrop to understand and interpret teachers' arguments and beliefs about ethical issues in Biological Sciences.

A background of ethical theories that would be used within the ethical matrix can be taken as follows (refer to Table 2.1).

2.2.1 Animal Experimentation

Experimentation on animals is known since ancient times, dating to around 500 B.C. were the older records of real anatomical observations. Studies on animals were also a central part of Aristotle's work (384–322 BC), as he is believed to have dissected over 50 species of animals. Some of the path-breaking researches that can be attributed to animal experimentation include study of movement of heart and blood in animals by William Harvey (1578–1657); conditioning in dogs by Ivan Pavlov (1890); isolation of three forms of polio virus by Jonas Salk (1940), and many others.

Arguments In favour of Animal Experimentation

Advances in Medicine and Health Care (Therapeutic): Sigma Xi, the scientific research society, defends the use of non-human animals in biomedical research by citing what they take to be the enormous benefits of that research:

> Results from work with animals have led to understanding mechanisms of bodily function in humans, with substantial and tangible applications to medicine and surgery (e.g., antibiotics, imaging technologies, coronary bypass surgery, anti-cancer therapies), public health (e.g., nutrition, agriculture, immunization, toxicology and product safety)…

Table 2.1 Comparison of major ethical perspectives

Ethical theory	Morals/rules/duties	Virtues	Outcomes	Principles	Care
Other names	Deontological/ rights-based ethics	Virtue-based ethics	Consequentialist/utilitarian ethics	Principle-based ethics	Care-based ethics
Focus	Act	Agent	Consequence	Context	Power/ relationships
Description	Actions that are considered right or wrong independent of the consequences. One is obliged to fulfil one's duties	Attributes, dispositions or traits that enable us to act in certain ways that develop our human potential (for e.g. Honesty, trustworthiness, integrity, etc.)	Consequence of an action granted extreme importance and actions that bring about the greatest amount of benefit and happiness to the greatest number of people who are worth pursuing	Four principles that set the pillars of moral life (respect for persons/ autonomy, justice, beneficence (do good) and non-maleficence (do no harm)	Focuses upon the relationships and the underlying power structures within a situation

Taken from An Ethics Primer—lesson ideas and ethics background, © 2008, NWABR. Also available online at www.nwabr.org

Research with animals has made possible most of the advances in Medicine that we today take for granted. An end to animal research would mean an end to our best hope for finding treatments that still elude us.

Generation of Knowledge and Tracing Evolution Through Vivisection (Non-therapeutic): Most of the medical researches as well as those done in a lot number of animal research institutions torture or kill animals for the sake of knowing the anatomical as well as physiological characteristics of certain organisms. This is usually done as part of comparative studies at the graduate and postgraduate levels, and even in higher researches where such experiments are used to trace the lineage (evolutionary studies).

Animals as Easy Models for Drug Testing: It is better to test the effect of a drug on lab animals rather than doing directly on human beings. So far as insulin is concerned, it was only by experimentation on dogs that it came to be learnt that removal of something manufactured by the pancreas caused diabetes. In the continuing debate between experimentalists and champions of the rights of animals, the discovery of 'insulin' remains a shining example of the benefactions experimental animals have conferred upon man (LaFollette & Shanks, 1996). Besides, the use of animal models in testing of new vaccines, anti-toxins and cosmetics is unmatched (Parascandola, 1998).

Anti-Vivisectionists' Ethical Arguments

Do Animals Have RIGHTS? Is it morally justified to cause pain or harm to one set of animals in order to provide some kind of benefit to humans? Is it justified to devalue or denigrate the life of an animal as compared to that of Humans (Regan, 2014)? All such questions pester the conscience to give a second thought to the practice of animal experimentation as a method in research.

Species Barrier as a Deterrent for Extrapolation of Drug-Testing Experiments: Animals do not serve as appropriate models for medical testing of drugs and other invasive treatments, due to the difference in basic physiological and psychological make-up.

One major example of animal and human differences is that of heart research being done on animals, frequently dogs. There could be varied factors responsible for heart diseases in humans such as fatty diet, irregular lifestyle, smoking, drug consumption, lack of exercise, persisting stress or anxiety, etc., none of which can be replicated in an animal.[1]

The Underestimation of Human Harms: Many medicines that are not toxic for test animals prove to be highly toxic for human beings. A medical disaster, in the case of thalidomide research,[2] Zomax and DES which were all tested on animals and

[1]Refer to An Anti-vivisectionist Reply to pro-vivisectionists most common arguments, with a focus on Anti-vivisectionists Unmasked (Produced by Seriously Ill for Medical Research—SIMR).

[2]See 'The thalidomide story explained', *Down To Earth*, April 16–30, 2010.

judged safe, had devastating consequences for the people who used them. Animal testing wastes time, too, by leading researchers in the wrong direction.

Utility of Research: There has been a huge round of criticism by animal protectionists and animal rights groups for the utility of conducting experiments on animals, for say, testing a new eye shadow, does our society really need another eye shadow? The stakeholder cosmetic company is the only one deriving profit out of it, but should such a profit be desired by putting animal life at stake (Parascandola, 1998)?

Anti-Vivisectionist Organizations and Forums

Some scientists, social activists, ecologists and wildlife conservationists joined hands together to fight for animal rights, and initiated a number of concern forums and organizations such as PETA[3] and another U.K.-based National Anti-Vivisection Society (NAVS). These have helped in generating awareness of animal abuse and ill-treatment in our so-called modern scientific society and have also unveiled the deplorable state of animals in some of the highly acclaimed research labs in India and other countries. These attribute to the faulty scientific procedures used leading to wastage, poor laboratory practices and a lack of appropriate animal care. The animal rights groups prohibit any kind of experimentation on animals for whatever reasons unless done in the right of animal health and benefit. They bestow equal rights to animals as given to humans and insist that if an experiment cannot be carried out on human subjects then it cannot be done on animals as well (Parascandola, 1998).

Animal Welfare Laws: With the burgeoning rate of epidemic spread and newer strains of infectious diseases taking a toll on human lives, it has become indispensable rather ethical to conduct some research with animals in the right of human and animal well-being. But it is also incumbent upon the experimenters to care for animals and their intrinsic rights. Thus, a mid-way path that could be adopted is 'responsible experimentation' (Giridharan, Kumar, & Muthuswamy, 2000). This involves a commitment towards animal welfare and alleviation of their pain and suffering. This can be affected by the improvement of husbandry conditions by providing for their basic requirements like space, movement, food, social interactions and adopting the 3R principle of replacement, reduction and refinement.

Alternatives to Animal Experimentation

With the advancement in sciences and scientific procedures for experimentation, many new techniques have evolved to provide the same or even better results without animal experimentation. These include the in vitro and Bioinformatics tools, cell culture including preservation of primary cell lines, 3-D cell cultures,

[3]PETA or People for the Ethical Treatment of Animals, founded by Ingrid E. Newkirk in January 2000, is an organization that works towards educating the policymakers and the general public about abuse of animals in different spheres and aiming towards an understanding and promotion of animal rights and respecting them.

Table 2.2 Ethical matrix for animal experimentation

Stakeholders	Outcomes	Deontological (duty and rights perspective)	Virtue	Principles
Animals		Animals also possess the right to live as we humans	Care and concern for animals	Principle of non-maleficence: animals have a right not to be harmed by anyone
Researchers and scientists	Generation of knowledge Advances in medicine and health care	Principle of 3R, i.e. reduction, replacement and refinement	Good treatment of animals in labs	Principle of safety: extrapolation of results from animals to humans can pose harm
Research organizations	Models for drug testing		Eliminate torture and pain to animals	
Animal labs		Animal welfare rules and regulations	Treatment of animals in labs	Protection and care of animals

MIMIC, etc. that have proved to be more efficacious and less time- and resource-consuming as experimenting on live animals (Akbarsha, Zeeshan, & Meenakumari, 2010) (Table 2.2).

2.2.2 *Assisted Reproductive Techniques (ARTs)*

Assisted reproductive technology is a recent technology in the sphere of infertility and offers some hope to the infertile couples, bringing about happiness in their lives. There are a plethora of such techniques that assist in bringing about pregnancy and childbirth, such as IVF[4] (In Vitro Fertilization), ZIFT[5] (Zygote Intra-Fallopian Tube Transfer), GIFT[6] (Gamete Intra-Fallopian Tube Transfer),

[4]IVF or in vitro fertilization is a technique wherein the fertilization takes place outside the uterus, i.e. in vitro where the development of the embryo takes place for some time after which it is transferred into the uterus.

[5]ZIFT or Zygote Intra-Fallopian Tube Transfer is a technique wherein the zygote is directly transferred into the fallopian tubes of the uterus of the mother.

[6]GIFT or Gamete Intra-Fallopian Tube Transfer involves the transfer of freshly recovered ova and conditioned spermatozoa into the fallopian tubes (p. 2265, Encyclopaedia of Bioethics).

ICSI[7] (Intra-Cytoplasmic Sperm Injection) and many others. These techniques may require two parents or three parents (in case a surrogate mother is required or a gamete donor is needed). However, adopting any of these techniques has its own cost in terms of expenditure, physiological, psychological and emotional turmoil that a couple has to go through. Owing to the benefits that such technologies offer, there has been a spurt in their commercialization which has led to the flourishing of a whole 'reproductive market'. Especially, the technique of IVF is the most rampant one, as it has many success stories attached to it, although we can say that those which are usually highlighted. Some reports which compel us to think averse have been tabulated as under (Table 2.3).

Some of the important issues that have been raised by the news items relate to certain risk factors associated with the technology such as birth defects and congenital defects in the unborn child. Also the impact of technology on the society's socio-economic, cultural and moral structure can also be seen in terms of the commodification of reproductive tissues including gametic cells (donor sperms and ova) as well as organs (such as uterus of the surrogate mother).

The issue of when a meaningful human life begins has recently acquired significance, especially while doing discussion over IVF. The wastage of embryos during the procedure of IVF is beset with many ethical issues, the most premier one being a threat to human potential and human life. All those extra embryos that are being discarded or cryopreserved and later left to die are having human potential in them. Do we have any right to create or destroy human life just like that? Kantian principle (1724–1804) says that, "ends cannot justify the means", applying which would mean that to create one pregnancy, so many human embryos cannot be sacrificed (Mastroianni Jr., 2003).

Other hazards associated with this technology relate to the impact on woman's health undergoing this treatment as well as the extent of physical and emotional turmoil. These can be taken as follows:

Risk of Multiple Pregnancies: Overstimulation of the ovaries sometimes leads to the release of ova from more than one ovarian follicle, thus resulting in multiple fertilization in the fallopian tubes as well as at other extra-tubal areas (ectopic pregnancy). This can even risk the life of the mother and lead to premature and abnormal childbirths.

Ovarian Hyperstimulation (OHSS): This is another complication and a syndrome associated with the IVF treatment and leads to fluid accumulation in the extravascular compartments between the follicles and the ovary. The various symptoms associated with it include, weight gain, severe abdominal pain, nausea, vomiting, or diarrhoea, electrolyte imbalance and oliguria (Encyclopedia of Bioethics, 2273).

[7]ICSI or Intra-Cytoplasmic Sperm Injection is a technique wherein the spermatozoon is mechanically inserted into the oocyte (p. 2265, Encyclopaedia of Bioethics).

Table 2.3 Some recent reports related to ARTs

News headline *(with trimmed content)*	Source and date
Birth defects more common in IVF babies: Babies conceived through certain fertility treatment techniques are about one-third more likely to have a birth defect than babies conceived without any extra help from technology, according to a review of several dozen studies	Retrieved from American Scientist[a] on 6 April 2013
"**Children conceived by IVF and/or ICSI are at significantly increased risk for birth defects**, and there is no risk difference between children conceived by IVF and/or ICSI"	Retrieved from http://www.reuters.com/article/2012/04/20/us-ivf-idUSBRE83J03M20120420 on 6 April 2013
Four Million Test-Tube Babies and Counting (7 October 2010): IVF helping gay couples and unmarried women, not to mention older women, some past menopause, who have pushed conventional biological boundaries	Retrieved from http://www.theatlantic.com/technology/archive/2010/10/four-million-test-tube-babies-and-counting/64198/ on 6 April 2013
Test-Tube Babies May Face Greater Health Risks Than Naturally Conceived Children (23 February 2010): Research shows that children born with the help of reproductive technology might be at a higher risk for genetic defects leading to chronic disorders	Retrieved from Scientific American available at http://www.scientificamerican.com/article.cfm?id=assisted-reproduction-genetics on 6 April 2013
Rhinoceros Undergoes Assisted Reproduction to Rescue Species from Extinction (30 May 2012): Conservationists are taking heroic measures to restore fertility to a three-footed Sumatran rhino, but it is unclear if this is a prudent undertaking	Retrieved from Scientific American at http://www.scientificamerican.com/search/?q=Rhinoceros+Undergoes+Assisted+Reproduction+to+Rescue+Species+from+Extinction+%28May+30%2C+2012%29%3A+&x=1&y=11 on 6 April 2013
For Sale: Human Eggs Become a Research Commodity (17 November 2009): Paying a woman for her eggs to use in stem-cell research has been a bioethical no–no for years. The decision, which allows payment of up to $10,000, will likely jump-start donations—and thereby research. Many bioethicists, however, worry that the financial incentive could exploit women and compromise their health	Retrieved from Scientific American at http://wwwscientificamerican.com/article.cfm?id=shelling-out-for-eggs on 6 April 2013
Millions of embryos created for IVF in UK "thrown away unused" (31 December 2012): Millions of human embryos created for In Vitro Fertilization (IVF) pregnancies in the UK are being thrown away unused, according to official figures. As a result, 93% of all embryos created—more than 3.3 million in all—are never used to generate a pregnancy	Retrieved from The Hindu available at http://www.thehindu.com/news/international/millions-of-embryos-created-for-ivf-in-uk-thrown-away-unused/article4258699.ece on 6 April 2013

(continued)

Table 2.3 (continued)

News headline *(with trimmed content)*	Source and date
Transparency is the victim, 27 January 2013: In the absence of a strict law regarding assisted reproductive technologies, various medical institutions in Delhi are flouting medical and ethical rules with aplomb, reveals a new study. India has emerged as the surrogacy outsourcing capital of the world	Retrieved from The Hindu on 6 April 2013

[a]American Scientist available at http://www.americanscientist.org/science/pub/birth-defects-more-common-in-ivf-babies also visit http://www.reuters.com/article/2012/04/20/us-ivf-idUSBRE83J03M20120420

Surrogacy and Law: Often, the surrogate mothers are being ill-treated and exploited financially by some profit-making organizations that refuse to pay them their promised amount. Citizenship issues also arise especially when foreigners come to India which is considered to be the cheapest surrogacy hub, then it so happened in the case of German twins who could not be granted any citizenship of either India or Germany where surrogacy is banned.[8] This is all due to the inadequate and loosely drawn legal contracts despite the existing Regulatory Bill on the Assisted Reproductive Technologies (*The Assisted Reproductive Technologies (Regulation) Bill*, 2010).

Feminists who argue against the use of ARTs (Assisted Reproductive Techniques) regard it as debilitating both woman's health and dignity, reducing them to mere "foetal containers". According to them, the desire for motherhood is a socially constructed phenomenon that has been forced upon them by the predominantly male medical establishment that uses women as 'living laboratories' (Bartholet, 1992; Rowland, 1992). In contrast, the other group of feminists regard the use of this new technology as providing women with multiple options for bearing children and can help in circumventing the problem of infertility faced by many couples (Macklin, 1994).

Adoption is also considered as a viable and more rewarding option for some but is fraught with problems such as availability of healthy children for adoption, long interval of wait, etc. plus some of the benefits associated with these ARTs cannot be replaced by adoption such as continuation of one's genetic kinship, experiencing a range of fulfilling experiences by the mother such as those associated with pregnancy and childbirth.

Thus, an ethical practice in this direction would ensure a careful monitoring of all these risk factors involved in the treatment and weighing appropriately the benefits and harms, so as to reach a balanced decision that should not sacrifice the health and well-being of the mother at any cost (Table 2.4).

[8]Refer to The Hindu Newspaper article, "The Business of Baby-Making" dated 28 September 2014.

Table 2.4 Ethical matrix for assisted reproductive techniques (ARTs)

Stakeholders	Outcomes (utilitarian)	Deontological (duty and rights perspective)	Virtue	Principles
Gamete donors (male/ female)	Even if any one of the parents is unable to produce viable gametes then a donor parent can donate his/ her gametes			Principle of beneficence: all those involved in the process of assisted reproduction should be benefitted by it
Parent Mother	The mother is able to conceive her own child with assisted reproduction and gets rid of the socially constructed disease of 'infertility'	It is within the reproductive right of the mother to go for IVF and produce her own offspring		Principle of non-maleficence: these techniques should not cause any physical, psychological or emotional harm, especially to the mother undergoing such treatment
Parent Father	Father is equally benefited by IVF			
Surrogate Mother	Surrogate mother is well-paid for lending her womb and giving birth to a child that is not genetically related to her	Surrogate mother does not hold any right on the child		
IVF Child	Child is born in the world and acquires a human potential Child may or may not be normal	Relationship of the child with Surrogate mother		

2.2.3 *Amniocentesis and Sex-Determination*

Amniocentesis is a diagnostic technique, wherein foetal cells extracted from the amniotic fluid are tested for the presence of any chromosomal disorder such as Down's syndrome, etc. However, the technique is being misused for the detection of X or Y sex chromosome, and if found to be a female foetus leads to abortion. This kind of female feticide has been rampant in India as well as other developing countries like China, etc. Another risk associated with the technique lies in harming the environment of the womb, and in rare cases leading to even miscarriages.[9] Therefore, to avert such eventualities, researchers have devised another method to determine the child's genetic sequence, by a sampling of mother's blood and testing

[9]Science News Daily, 4 November 2006.

it for the presence of foetus cells in the second trimester.[10] However, despite its side effects the technique of amniocentesis is rampant in patriarchal societies such as India, leading to a steep decline in sex ratio (In India, it dropped from 927 females (for every 1000 males) in 1961 to 919 females per thousand males in 2011).[11]

The ratio in urban areas is even worse as compared to the rural sector. This has led to increased incidence of trafficking of poor women and forced polyandry in some regions with markedly skewed ratios.[12] The root cause behind such malpractices and the misuse of medical technology is the prevailing patriarchy and inequality of gender in the society at large. The constitution's amendment in this regard for bringing about an egalitarian social order such as prenatal diagnostic techniques act of 1994 was also not effectively implemented. The act was also amended in 2003 in order to include the preconception sex-selection techniques within its ambit, thus also ensuring a tighter control over the use of the technology, registration of diagnostic centres as well as monitoring of the medical personnel involved. Advocates of this technology of prenatal genetic screening whether it is via amniocentesis, chorionic villus sampling (CVS) or any other method such as ultrasound, etc. state that this would give women their liberty to choose the offspring, whether in terms of sex-selection and sex-selective abortions or selection of a healthier offspring (referred to as reproductive autonomy). There are a plethora of ethical issues that surmount such pre-implantation genetic diagnostic technologies, some of which can be discussed as follows.

Social versus Medical Reasons: Sex-selection techniques are sometimes used for the detection of any sex-linked genetic disorder which cannot be treated under the presently available medical treatment. In such cases, the techniques such as amniocentesis or any other PGD (Pre-implantation Genetic Diagnostic) technique is justified and morally acceptable. But, for reasons other than this such as preference of one sex over the other, for one's own subjective reasons is highly unacceptable.

Personal Autonomy and Reproductive Rights: One of the major arguments that go in favour of such sex-selection techniques is the practicing of one's reproductive autonomy and an open choice to select the sex of the child if the techniques are readily available and accessible too. This kind of liberty for procreation is being offered in western democracies, as it in no way is going to harm anyone else,

---

[10]Popular Science by Rebecca Boyle posted 6 July 12 retrieved from http://www.popsci.com/science/article/2012-06/new-breakthrough-scientists-decipher-almost-entire-genome-unborn-baby on 5 April 2013.

[11]Source: http://pib.nic.in/newsite/PrintRelease.aspx?relid=103437.

[12]Sex ratio, patriarchy and ethics retrieved from http://www.thehindu.com/opinion/lead/sex-ratio-patriarchy-and-ethics/article1819313.ece on 5 April 2013.

especially acceptable in the case of sperm sorting[13] techniques. According to some blocking access to sex-selection technologies, especially for those who want a child of a particular sex can be taken as a severe curtailment of their reproductive autonomy (Encyclopedia of Bioethics, 3rd edition, pp. 22–69).

Treating Children as Persons: According to the Kantian principle, "People should not be treated as a means to some ends rather ends in themselves"; therefore, if a child of a particular sex is being selected then this would violate the process of randomness associated with the natural process of sex-selection during embryo development. This is not consistent with respecting their children as persons in themselves (Encyclopedia of Bioethics, 3rd edition, pp. 22–69).

Gender Equity: A balanced sex ratio is an indication of a healthy and gender-neutral society, whereas a society in which preference is shown towards a particular sex is regarded as unequal societies, where equality is denied. In societies where males are given a greater preference over females are indicative of inter-social warfare, greater frequency of violence and violence against women.

However, this argument spurs another one which says that these technologies are promoting eugenics or selection of best traits in organisms, but who would decide which are those best traits? (Encyclopedia of Bioethics, 3rd edition, vol. 3, pp. 996–1003).

Sociopolitical angle: Banning such a useful technique will not serve the purpose, as some unscrupulous elements would still find the way out; moreover, there would be increased incidence of chromosomal disorders such as Down's syndrome amongst newborn children that can hamper the overall health and development of the society. The fault does not lie in the medical technology per se, rather in the mindset and psyche of the population in the majority that is diseased and irrational. The bigger debate here is between women liberty, her right to live a safe and secure life and the gender-biased patriarchal society that favours the male child. Women are considered to be a 'burden' on the family income, as more has to be spent at the time of marriage as dowry and lesser returns; therefore, people prefer to go for feticide in the case of a girl child as quoted by one, "better 500 now than 50,000 later."[14] The following news items reveal the state of greed and craving for the male child.

A survey report of women's centre in Mumbai found that out of 8,000 foetuses aborted in six city hospitals, 7,999 foetuses were of girls (Gangrade, 1988: 63–70). It is reported that about 4,000 female babies are aborted in Tamil Nadu (southern India) every year.

---

[13]Sperm sorting is a medical technique for sorting out sperms carrying X or Y chromosome and then the selected sperm is fused with the ovum giving rise to an embryo of desired sex.

[14]The Other Half—Where have all the girls gone? By Kalpana Sharma published in The Hindu, dated: 16 April 2011, retrieved from http://www.thehindu.com/opinion/columns/Kalpana_Sharma/the-other-half-where-have-all-the-girls-gone/article1701575.ece on 5 April 13.

Table 2.5 Ethical matrix for amniocentesis and sex-determination

Stakeholders	Consequentialist	Deontological (duty and rights perspective)	Virtue	Care	Principles
Females	Steep decline in sex ratio of females	Right to life		Care for the girl child	
Males	Patriarchy		Equal opportunities to all	Care for the female members	No harm to be posed to female foetus
Society	Skewed sex ratio, human trafficking, forced polyandry, etc.	Ban on amniocentesis	Gender equality	Care for every member of the society	
Medical professionals			Non-revelation of sex of the foetus in diagnosis		
At-risk population	Diagnosis of inherited and sex-linked disorders				

In Punjab and Haryana, these sex-selective abortions increased from 62,000 to 69,000 in Haryana and from 51,000 to 57,000 in Punjab (The Hindu, 19 October 2001).

Thus, there is a greater need for promoting gender equality not just in the documents but also in practice by debating and discussing on the prevalent socio-cultural practices prevailing in our society that tend to subvert the female rising and deny them even the basic human rights. The issue needs to be dealt with a greater sensitivity and seriousness, as skewed sex ratio can germinate some grave societal evils and sex-linked crimes, such as polyandry, increase in rapes, human trafficking, etc. (Table 2.5).

2.2.4 Genetically Modified (GM) Crops

GM crops involve the insertion, deletion or silencing of genes to give rise to a new organism with modified characteristics. These modified traits in plants can include pest resistance, enhanced water-retention capacity, better yield, increased shelf life, added nutrient value, etc.

Arguments in Favour of GM Crops

The major objective behind the introduction of GM Technology, in India particularly, has been multifold, including the concerns of hunger, poverty and economic trade-off. These have been discussed as follows.

Issue of Hunger: The added qualities and features of GM crops such as pest resistance and drought resistance are regarded as antidote to solve the problem of hunger of many countries including India.[15] The conventional agricultural methods of crop production do not offer much promise in eradicating the problem.

Issue of Poverty: GM technology offers an incentive to the poor farmers, by increasing the crop yield, offering climate resilience, as well as profit-making, thus annihilate their agony and dependence on inadequate rains and extra funds to raise the crop. The cost of production also reduces on account of lesser need for pesticides in GM crops. The number of applications of pesticides in non-Bt. crops was 19.8 as against 6.6 in Bt. cotton. The cost of pesticide use per hectare came to $726 and $136 for non-Bt and Bt. cotton, respectively.[16]

Issue of Acreage and Habitat Generation: It is now a well-known fact that GM crops require lesser land area as compared to the non-GM crops, which saves the rest of the area for forestry and habitat for the wilderly. Norman Bortang, associated with the Green Revolution, has also postulated that the new technology—biotechnology—should be allowed to advance in the welfare of human race so that the vast stretches of areas can be reverted to forest and wildlife habitats.[17]

Added Nutrient Value and Healthier Foods: GM food is said to have equivalent nutritional value as the conventional non-GM food, more so crops with added nutrient value can be produced, such as GM Golden Rice having additional vitamin A. Besides, due to lesser usage and application of pesticides, the food produced is healthier.

Advantage of Trait Selection: The principles of Biotechnology involved in the creation of GM crops help in the selective transmission of only selected traits in a controlled and sophisticated way that is not a possibility with conventional methods.

Advantage of Combating Abiotic Stresses: This GM technology has enabled in the production of crops that are herbicide tolerant, insect/pest resistant, bacterial/fungal/viral resistant and able to thrive in extreme climatic conditions such as drought,

---

[15]As per the Global Hunger Index 2018 report India ranks 103rd out of 119 countries that participated. (https://www.globalhungerindex.org/results/).

[16]Refer to Khosla, P. K. (2002), "Eco-friendly Bt. cotton and GMCs saviour of Indian farmers" in *Agriculture Tribune*, Monday, 20 May 2002, Chandigarh, India.

[17]Refer to Das, N. M. (2006, 2 January). Food Security through Genetic Engineering. The Hindu. Retrieved from http://www.hindu.com/edu/2006/01/02/stories/2006010200410200.htm.

unfavourable soil conditions, etc.[18] In developing countries, many resources are being spent to overcome all such vicissitudes to increase the crop productivity; hence, GM crops offer a ray of hope and provide stability and high yield to poor farmers.

Generating Employment: With growing population and increasing unemployment, the agriculture sector is a hope as it is less capital-intensive and requires lesser equipment and machinery as compared to the urban sector. An argument given in favour of GM crop plantation is that it would increase the crop yield which would intensify the recruitment of labour on farms and hence increased opportunities for employability and profit to poor farmers, especially in developing countries.[19]

Arguments Against GM Crops

Recent researches with regard to GM crops, their production, processing as well as field trials have raised serious doubts and concerns in the minds of scientists, environmentalists and the major stakeholders, i.e. consumers.

Food Quality and Nutrition: Genetic modification of plants may result in alteration in the nutritional profile of the plant product which can also result in altering the nutritional status of the consumer. This can result in nutrient imbalance in the body as well as impact the overall dietary intake (FAO Corporate document repository, 2000). Currently developed plants with improved nutritive value include GM rice with enriched vitamin A and GM soya bean and rapeseed with modified fatty acid.

Food Safety: The inserted gene, or even the insertion process itself, may re-engineer the biology of the plant and generate poisons or toxins.[20] The case of GM potatoes experiencing *Galanthus nivalis* lectin gene for insecticidal properties is an example of the potential of GM foods to cause toxicity. For example, in a group of rats fed with GM potato caused damage to their immune systems and stunted growth and the experiment had generated considerable controversy (ICMR, 2004).

Gene Pollution and Loss of Biodiversity: The contamination caused due to GM varieties of plants through pollen drift can cause loss of biodiversity. This was the reason behind the disapproval of GM corn variety for commercialization in Mexico, as the native corn varieties might get contaminated by the foreign genes.[21]

---

[18]From "The use of genetically modified crops in developing countries—a follow-up Discussion Paper", Nuffield Council on Bioethics, © 2003 downloaded from http://www.conacyt.gob.mx/cibiogem/images/cibiogem/comunicacion/publicaciones/Nuffield_Council-GMOs-for-dev-countries.pdf on 14 June 2014.

[19]From "The use of genetically modified crops in developing countries- a follow-up Discussion Paper", Nuffield Council on Bioethics, © 2003.

[20]Refer to Purkayastha, P. and Rath, S. (2010, May 15). Bt Brinjal: Need to Refocus the Debate. *Economic and Political Weekly XLV* (20), 42–48.

[21]Anuradha, R. V. (2002). GMOs—Promises and Concerns. Frontline, Volume 19, Issue 08, April 13–26, 2002.

Antibiotic Resistance—Potential for Gene Transfer: Sometimes, it has also been reported that GM food (particularly Bt-Brinjal) may lead to the generation of an antibiotic-resistant protein leading to alteration in blood chemistry including blood clotting time (prothrombin), total bilirubin (liver health) and alkaline phosphate in goats and rabbits.[22]

Emergence of Superweeds: As per some recent reports on GM field studies, the herbicide-resistant gene that is being genetically transferred to the GM crop can sometimes cross the species barrier and get integrated with the genome of some wild relatives of GM crops, which then become resistant to the effect of pesticides or herbicides.

Mixing of Genes—"Breach of Religious Faith": Many religions have explicit dietary prohibitions against certain foods or consuming particular foods (Pascalev, 2003; Knight, 2009). Consensus conferences in Australia and the United Kingdom highlight lay concerns about mixing human and animal DNA with plants, which is seen as tantamount to cannibalism.

Agricultural Knowledge Dissonance Leading to an Upsurge in Farmer Suicides: This is not to say that GM seeds are the sole cause of farmer's suicide, one of them, and can be explained on account of agricultural knowledge dissonance.[23] This is increasingly leading towards rejection of indigenous methods of production, and following the suite of developed nations by adopting advanced agri-based technology. Thus, all of these factors contribute towards farmer's distress (Table 2.6).

2.2.5 Contraceptives and Birth Control

A safe, reliable and easy to use contraceptive has always been the need of the hour in order to curb the growing population as well as to provide resistance against a large number of Sexually Transmitted Diseases (STDs). Earlier, the methods used for contraception were restricted to periodic abstinence and withdrawal, which were more of the natural kind. The nineteenth and twentieth centuries saw a spurt in research in the area of contraceptives, as the natural methods were less reliable and resulted in unwanted as well as untimely pregnancies. Thus, coming up of an era of artificial contraceptives, that can be categorized as under.

---

[22]Refer to "India says No to Bt-Brinjal", retrieved from http://www.bhoomimatha.com/india-says-no-to-bt-brinjal/.

[23]Rejection of indigenous methods of production, and following the suite of developed nations by adopting advanced agri-based technology, whether they comply with Indian climatology and topography or not. See Vaisavi, A. R. (2004). Suicides and the Making of India's Agrarian Distress. National Institute of Advanced Studies, IISc Campus, Bangalore, India.

Table 2.6 Ethical matrix for genetically modified (GM) crops

Stakeholders	Consequentialist utilitarian	Deontological (duty and rights perspective)	Virtue	Care	Principles
Farmers	Productivity, economics, food security	GM crops versus organically grown crops			Farmer autonomy
Agricultural scientists	Patents, publications	Biopiracy	Fair reporting of results/ trials		
Seed companies (Indian and Multinational)	Profits, distribution of quality seeds	Farmers under debts, sign contracts			Justice to farmers
Crops (natural wild variety)	Horizontal gene transfer, dilution of gene pool	Threat to species diversity			
Crops (genetically modified)	Health hazards, allergies, superweeds	GM crops should be tested well before their plantation in the soil			Non-maleficence: environment should not be harmed

- Barrier methods (using male and female condoms, cervical caps, diaphragms, etc.),
- Chemical methods (including Oral contraceptives/hormonal pills, emergency contraceptives or i-pills, IUDs, skin-implants, spermicidal jellies and creams, etc.) and
- Surgical methods (Vasectomy in males and Tubal ligation in females).

Of these, the surgical methods offer irreversible means of contraception, whereas the others can be used reversibly.

Socio-ethical Debates Surrounding the Use of Contraceptives

The advocates of all these birth control methods have always been of the view that use of such contraceptives will secure the liberty and right of the woman on her own body and would provide her with a choice. As, earlier, when such contraceptives were not available or were only accessible to a few richer sections of the society, the women were forced into abortions as they could not bear the burden of an extra child. In the dearth of adequate medical facilities and prenatal care, the women forced themselves into abortion by using needles, etc. that many a time led to septicaemia resulting in the death of so many mothers. On the other hand, those from the rich and elite families had better knowledge as well as access to these artificial contraceptives and hence used them to curb their family size as well as save health. This kind of an unequal distribution of knowledge and means was for the first time noticed by an Irish lady named Margaret Sanger, who devoted her entire life towards the cause of making contraceptives available to especially the women from the poorer sections of the society, along with the right kind of sex education to the adolescents about the use of these contraceptives. However, her position was being condemned by some as having 'eugenic' intention and a racist ideology because this would limit the number of immigrant, poor, uneducated and deprived population. However, some minority groups have urged for the advancement and distribution of contraceptives owing to the increased incidence of unintended pregnancies and rise in the number of STDs, thus calling for a culturally sensitive family planning and contraception services (Encyclopedia of Bioethics, pp. 902–903).

Pro-family Versus Welfare Conservationists

Wherein the pro-family movement regards the use of contraceptives as going against the sexual morality evidenced by high rates of abortion, rise in the number of adolescent and teenage pregnancies, out-of-wedlock births, and sexually transmitted diseases. In contrast, the welfare conservatives regard the women health at priority and are concerned about the declining sexual health of women owing to frequent abortions, illegitimate births, teenage and adolescent pregnancies, etc. (Encyclopedia of Bioethics, p. 906).

Impact of Contraceptives on Women's Health

Some contraceptive pills, especially those that are ingested (oral contraceptives or OCs) or implants and hormone injections, have some beneficial as well as adverse impacts on those using them. Some of the beneficial effects of OCs include a lowered risk of salpingitis, a 50% reduction in the risk of the development of both the endometrial and ovarian cancer, as an excellent therapy for primary dysmenorrhoea, prevention against ovarian cysts, etc. Some of the adverse effects of OCs include nausea, weight gain, heavy bleeding between menstrual periods, headaches, etc. (Table 2.7).

2.2.6 Genetically Modified Organisms (GMOs)

Genetically modified or transgenic organisms are being created by inserting or modifying the given genetic composition of any organism so as to enhance its vigour or quality or simply inserting desired characteristics. These organisms can include plants, microbes, viruses or other animals; some examples are GM cotton, genetically engineered insulin produced by a genetically modified bacteria, bacteria designed for the treatment of oil spills in rivers and oceans, etc.; however, the new buzzword associated with these transgenic organisms these days is 'Synthetic Biology' or 'Biodesign'.[24] Transgenic animals are also being used in identifying a

[24]For more information visit http://www.bbsrc.ac.uk/web/FILES/Reviews/0806_synthetic_biology.pdf.

Table 2.7 Ethical matrix for contraceptives and birth control

Stakeholders	Consequentialist utilitarian	Deontological (duty and rights perspective)	Virtue	Care	Principles
Women in reproductive age bracket	Reproductive freedom	Reproductive right and choice			Non-maleficence: woman's physical and emotional health
Pharmaceutical companies	Profit-making	New and improved contraceptives			Non-maleficence: minimal side effects

particular gene's function by tagging it with a fluorescent marker gene, for example, a GFP gene that gives rise to a fluorescent protein in jelly fish. Another important application of transgenic animals is in the synthesis of pharmaceuticals and some rare proteins or hormones referred to as "Biopharming". This includes the synthesis of transgenic insulin and vaccines which can be extracted from the milk of animals. An important anticoagulant drug (A Tryn) has been developed by using the same method by extracting it from goat's milk.[25] Similarly, another protein extracted from transgenic goat called as 'spider silk' has a huge potential ranging from medical sutures to body armour.[26]

Although having numerous benefits, genetically modified organisms have some inherent issues of concern that need to be addressed such as follows:

Release of GMOs in the Environment: There should be a proper means and procedure for the release of these GMOs into the environment. Certain mechanisms such as engineering them to become dependent on certain nutrients with limited availability or automating with some self-destruction mechanism to get triggered once their population size increases[27] can be devised.

Threat to 'Bioterrorism': Synthetic organisms can play havoc with the environment, especially because they are designed on human will. The synthesis of polio virus (Josefson, 2002) and the Pandemic Spanish Flu virus of 1918 (Tumpey et al., 2005) by the scientists in order to study the virulence patterns of these viruses are some such examples of creation of GMOs. This raises serious concerns over the intentions of those designing organisms for their own interests such as some maverick technologists who may use it to spread bioterrorism and bio-weapons (see footnote 27). Therefore, norms have to be established along with proper judicial

---

[25]Chapter-2 on 'Biotechnology Promises and Expectations' downloaded from http://www.nuffieldbioethics.org/emerging-biotechnologies/emerging-biotechnologies-biotechnology-promises-and-expectations on 23 May 2013.

[26]Lazaris, A., Arcidiacono, S., Huang, Y. et al. (2002) Spider silk fibers spun from soluble recombinant silk produced in mammalian cells *Science* 295: 472–6.

[27]Refer to http://www.bbsrc.ac.uk/web/FILES/Reviews/0806_synthetic_biology.pdf.

Table 2.8 Ethical matrix for genetically modified organisms (GMOs)

Stakeholders	Consequentialist utilitarian	Deontological (duty and rights perspective)	Virtue	Care	Principles
GMOs	Pharma factories Biopharming	Informed consent Physical harm and distress to animals Intrinsic value of animals			Non-maleficence: bioterrorism
Scientists and research organizations	Biopatents' Growth in research and development		Not to conduct research on GMOs for profit-making	Care for the environment	GMOs should not pose any harm
Society	Alternative sources of essential proteins, hormones at cheap cost	Labelling of GM food			
Environment	GMOs escape into the environment, developing resistance in native species	Preserving the integrity of the environment and species			

rules and regulations for keeping a check on the commercialization of the DNA synthesis and licencing should also be made mandatory (see footnote 27).

Bio-Patenting of GMOs: Certain useful GMOs such as insulin-producing bacteria, or the bacteria that can turn biomass into fuel (such as ethanol) if patented by the privatized companies, would lead to monopolization in the market, which would hinder the basic research and distribution of technology. Such a practice would create a divide in society by making the technology unavailable to the general populace (see footnote 27).

'Playing God' by Creating Artificial Life: In 2003, Craig Venter had announced that it had taken his team just 14 days to design the first synthetic virus from scratch, bacteriophage phi X 174. Venter's team has also been successful in their endeavour and published their first ever synthetic-bacterial genome in January 2008.[28] Venter's experiments have been opposed by many scientists and non-science people as playing with nature's creation and gene pool. Such artificially designed organisms by interacting with other organisms can cause alteration in a gene pool which could be devastating (Table 2.8) (see footnote 27).

[28]For more information visit http://www.jcvi.org/cms/research/projects/synthetic-bacterial-genome/press-release/.

2.2.7 Stem-Cell Therapy and Research

The identification and isolation of stem cells[29] opened up a whole new era of treatment and research in the field of Medical Biotechnology. Especially in the case of organ transplants where finding an appropriate donor is difficult and even if one finds it, the possibilities of organ rejection by the recipient's immune system may hinder such transplantations. Therefore, patient's own stem cells are sought for developing organs which will not be rejected and hence offer a perfect treatment. Till date, stem-cell treatment has been effective in treating macular dystrophy,[30] multiple sclerosis,[31] Parkinson's disease,[32] sickle-cell disease, anaemia and thalassaemia. Stem cells are also now used for the purpose of disease modelling and for the production of in vitro models for disease testing.[33] Stem cells are usually found in the bone marrow (bone marrow stem cells) and in the embryonic tissues (embryonic stem cells). However, the embryonic stem cells are known to have the maximum regenerative capacity as well as totipotency as compared to the bone marrow stem cells. Many ethical issues are raised when stem cells are being derived from the embryonic tissues or when embryos are grown specifically for the sake of deriving stem cells.

The Moral Status of the Embryo: This is already under a huge debate, as to whether an embryo should be treated as a person or at least a potential human being. Thus, if the human embryo is to be regarded as a potential human, then it would be difficult, rather impossible to treat it as a means to an end, as per the Kantian principle that human beings cannot and should not be regarded as a means to an end rather an end in themselves.[34]

However, in order to put an end to this ethical dilemma, scientists have devised a new way of reprogramming of the normal skin cells and inducing them to behave like stem cells. This is done by way of triggering a certain set of genes 'On' so that the cell regains its embryonic properties and switching 'Off' its present function (Table 2.9).[35]

---

[29]Stem cells are the precursor cells which have the capacity to give rise to any specialized cell type of the body as well as the ability of self-renewal.

[30]Taken from http://www.guardian.co.uk/science/2011/sep/22/embryonic-stem-cell-trial-blindness?newsfeed=true on 9 April 2013.

[31]Refer to http://www.the-scientist.com/?articles.view/articleNo/32123/title/Could-Stem-Cells-Cure-MS-/.

[32]Taken from http://www.guardian.co.uk/science/2011/nov/06/stem-cells-brain-parkinsons-disease.

[33]For further information visit http://www.nuffieldbioethics.org/emerging-biotechnologies/emerging-biotechnologies-biotechnology-promises-and-expectations.

[34]Refer to Stem-cell Therapy: the ethical issues, a discussion paper, © Nuffield Council of Bioethics 2000 or download from web: bioethics@nuffieldfoundation.org.

[35]Refer to http://www.guardian.co.uk/science/2009/mar/01/stem-cells-breakthrough.

Table 2.9 Ethical matrix for stem-cell therapy and research

Stakeholders	Consequentialist utilitarian	Deontological (duty and rights perspective)	Virtue	Care	Principles
Embryos (from where stem cells are derived)	Human embryos discarded after taking out stem cells Wastage of human embryos and human potential	Informed consent Violation of human rights	Avoid wastage of human embryos		
Medical and research institutes (offering stem-cell treatment)	Stem-cell treatment for previously untreatable disorders such as Alzheimer's, Parkinson's, etc.	Accessibility and affordability of treatment Sacrificing human embryos			Justice: Research carried out on some selected diseases

Issue of Informed Consent: A proper consent should be taken from the donor parents from whom embryos are being taken for carrying out research. The implications of that particular research should also be told to the parents, so as to make them aware of what use their embryos will be put to. Also, an assurance should be given to them that the embryos will not be implanted into any surrogate mother's uterus and would be used only for carrying out research (Scully, Haimes, Mitzkat, Porz, & Rehmann-Sutter, 2012).

Principle of Sentience and Human Potential: There have been so many debates around the status of human potential inherent in human embryos. Killing the foetus for the sake of stem cells is a serious loss to the foetus as it deprives it of its future valuable life.[36]

2.2.8 Human Genomic Project and Gene Therapy

Human Genome Project is one of the most fascinating scientific projects of the twentieth century spearheading a whole new arena in the field of medicine and therapeutics. The primary aim of the project was to determine the sequence of base pairs that make up the whole DNA of humans. Along with this, the project also aimed at finding out the function of different genes that are constituted by a DNA segment. A 13-year effort, which began formally in 1990 and got completed in

[36]Marquis, D. (2008). Why abortion is immoral. In Contemporary issues in bioethics, 7th edition, ed. T. L. Beauchamp, L. Walters, J. P. Kahn, and A. C. Mastroianni. Belmont: Thomson Wadsworth.

2003, led to the discovery of around 20,000–25,000 human genes and made them accessible for further Biological study and research.[37] The chief goals of the project were

1. To identify all the genes of the human genome (initially estimated to be 100,000),
2. To get the correct sequence of the human genome,
3. To develop databases to store this information,
4. To develop tools for data analysis,
5. To address the ethical, legal and social issues,
6. To sequence thc genome of model organisms, such as E.coli, Drosophila, etc.[38]

All of these goals have been successfully achieved by the scientists; however, one of the major concerns is the goal 5 which purports to address the ethical issues related to this technology. This led to the establishment of the joint working group that takes into consideration the ethical, legal and social implications of human genome research. The group was chaired by Nancy Wexler and held its first meeting in September 1989. The four major areas that were kept at high priority by this group are as follows:

- Quality and access in the use of genetic tests.
- Fair use of genetic information by employers and insurers.
- Privacy and confidentiality of genetic information.
- Public and professional education (Wexler in Cooper, 1994, p. 321).

The major applications of HGP include those in the sphere of medicine and therapeutics, especially with some of the initially thought to be incurable diseases, such as SCID (Severe combined Immunodeficiency) syndrome, deficiency or augmentation of any particular gene or protein defect. These defects are usually treated by the way of somatic-cell gene therapy that replaces a defective gene with its functional copy or augmentation of a deficient gene or protein. One of the marvels of this technology and the treatment offered with its help is the treatment of autosomal recessive SCID disorder in 'Ashanti De Silva' who received the gene therapy treatment under the approved protocol for the first time and is since then leading a normal life.[39] But, on the other hand, there had been some major setbacks including the death of Gelsinger, a patient who received gene therapy via adenoviral vector and died due to severe immunological response in 1999.[40] This incident and many other unsuccessful gene therapy trials have raised serious

[37]Retrieved from http://www.ornl.gov/sci/techresources/Human_Genome/project/hgp.shtml on 28 May 2013.

[38]Stanford Encyclopaedia of Bioethics.

[39]"Ethical Issues in Human Gene Therapy" Taken from http://www.ornl.gov/sci/techresources/Human_Genome/publicat/hgn/v10n1/16walter.shtml.

[40]Hunt, S. Y. (2008). Controversies in treatment approaches: Gene therapy, IVF, Stem cells and Pharmacogenomics. Retrieved from http://www.nature.com/scitable/topicpage/controversies-in-treatment-approaches-gene-therapy-ivf-792.

concerns and apprehensions over the reliability and efficacy of this procedure and at the same time developing more sophisticated methods of gene delivery and the effect thereafter. Some important areas of ethical dilemma and conflict with regard to human genome project and gene therapy can be taken as follows:

- Favourable risk–benefit balance (principle of beneficence/non-maleficence).
- Informed consent (principle of respect for persons).
- Fairness in selecting research subjects (principle of justice).

Gene Therapy Versus Enhancement: Gene therapy is ethically acceptable till the objective is the treatment or correction of a particular disease or disorder, but, when it comes to modifying or enhancing a given normal human characteristic or trait such as intelligence, beauty, athletic abilities and any other such trait, then their acceptance is liable to ethical scrutiny.[41]

Somatic-Cell Therapy versus Germ Line Gene Therapy: Somatic-cell gene therapy entails the correction or replacement of genes present in the somatic cells of the person or embryo, whereas germ line gene therapy causes modification in the germ cells which means that a particular gene therapy treatment would be carried over to generations and would be inheritable. Although such an inheritable treatment would provide a permanent cure for a particular hereditary disorder and would save the coming generations to undergo such a treatment again and again, tinkering with the germ line cells is beset with many ethical issues.

Reliability of the Technique and Insertion Process: The major ethical issue raised is that if the inserted gene gets misrepresented in the germ line then it would be devastating for all the generations to come (Fletcher, 1983). Or, there could also be a possibility of some tumour-inducing genes to get activated by the insertion or deletion of a particular gene, which is the risk factor equally associated with the somatic-cell and germ line gene therapy.

Genetic Reductionism: It is usually being said that we are our genes and the same has been advocated by Watson, the director of the HGP, when he said:

> The Human Genome Project is much more than a vast roll call of As, Ts, Gs, and Cs: it is as precious a body of knowledge as humankind will ever acquire, with a potential to speak to our most basic philosophical questions about human nature, for purposes of good and mischief alike (with Berry 2003, p. 172). (Gannett, 2010)

This statement of his clearly indicates the prospects of HGP as a journey into the unknown world of human nature, its origin, and how it can be modified, which has both sides of the coin, such that it can be used for the good and for the detriment of human species. But the question over here is that whether genes are the sole controller of a person's behaviour and have a deep impact over self-identity and self-understanding? Therefore, can human beings be reduced to a 'beanbag' of

[41]Ethical Aspects of Gene Therapy by Alex Mauron retrieved from http://www.gfmer.ch/Endo/Lectures_09/ethical_aspects_of_gene_therapy.htm.

genes? If the answer is towards the affirmative, then this would mean that by altering certain genes, we can modify or eliminate any particular trait or behaviour. This would then give rise to the whole set of pre-implantation genetic diagnostic techniques (PGD) and genetic engineering of embryos by selecting the most suitable genes for one's offspring, which would spur many ethical conflicts related to genetic determinism and the nature–nurture debate (Gannett, 2010).

Gene Patents and Biopiracy: Granting patent is a way of acknowledging and rewarding the person who has discovered or synthesized a novel product and anyone who wants that product has to take permission from the person to whom the patent has been granted, and has to pay some royalty as well. Earlier only those products which were the outcome of individual effort and design were granted patents, but a huge controversy arose when the natural products such as genes or protein that were although synthesized artificially in the lab were also being filed by the scientists claiming for patent rights. Some of the Western research organizations have patented the medicinal and food uses of plants and their products which have long been part of the traditional indigenous knowledge and can be taken as instances of 'Bioprospecting' and 'biopiracy' (Shiva, 1996). Such practices need to be checked and the indigenous resources need to be preserved. There have been many oppositions to the act of patenting, as it would lead to building secrecy and not allowing sharing of research amongst the scientists; moreover, it would engage scientists in short-term projects aimed at minting profits while disregarding the human ethos and values especially in the patenting of human genes whether natural or artificial (Nelkin, 2002) (Table 2.10).

Table 2.10 Ethical matrix for human genomic project and gene therapy

Stakeholders	Consequentialist utilitarian	Deontological (duty and rights perspective)	Virtue	Care	Principles
Human society	Treatment of diseases 'Pharmacogenomics'	Right to better treatment Right to privacy of genetic information	Misuse by employers, insurance companies		Non-maleficence: safety of in vitro gene therapy procedures
Embryos (genetically engineered)	Genetically designed babies	Informed consent Manipulation with genes Genetic reductionism Divide in the society			Equality
Research scientists and medical institutes	Patents and Biopiracy	Patenting of naturally occurring genes			Justice

2.2.9 Biodiversity Conservation

Biodiversity can be defined as the large variety of flora, fauna, soils and ecosystems present on this planet earth. It contributes to the inter-species as well as intra-species diversity and adds to the gene pool. This Biodiversity is of major importance for the sustainable development and growth of the human population as well as other organisms including plants and animals.

Biodiversity and its conservation has recently come on to the priority list of the countries world over due to the rapidly changing climatic conditions as well as human interference that have contributed to the extinction of species on this planet. A need has therefore emerged to control and curb this growing rate of species extirpation. In order to achieve this goal, many conservation policies (*The Wildlife (Protection) Act*, 1972) have been formulated, discussed and debated in conferences and conclaves but little concrete could be achieved. This is especially evident by looking at the recent list of endangered species of animals and plants published by IUCN (International Union for Conservation of Nature and Natural Resources) Red list. Hence, the conservation efforts worked out by different world policies need to be re-examined for the lacunas and appropriate steps need to be taken. The various issues that generate social, political, economic, moral and ethical concerns vis-à-vis conservation of Biodiversity can be listed as follows.

A Human–Mammal Conflict: In the areas that lie in close proximity with the forests or zones with rich biodiversity, there often occurs a conflict between the tribes and locals inhabiting that place and the biodiversity therein. This is especially with regard to the conservation policies that mostly favour the conservation of flora and fauna but fail to take any cognisance of the local communities whose only source for income generation or food, shelter and other important resources is being derived from these nature's reserves. These policies are usually framed by the top government officials who do not find it necessary to consult the people of the local communities who are going to be directly affected by the policy's implementation (Hill, 2002).

Conservation Versus Economic Growth: Often the conservation efforts tend to interfere with the economic goals of the country, and then the choice becomes difficult as to what should be preferred? And if it comes to developing countries, then the choice is more likely to be towards the economic and development side. An example that can be cited here is of timber industries and the associated logging that leads to falling of trees, hence the deforestation and habitat destruction for many species of animals. But often the two goals of conservation and economic growth conflate and coincide with each other, such that conservation of biodiversity promotes the economic growth of a country. The example that can be cited here is of the utility of animals and plants in generating fur, meat, hides, fruits, vegetables, cash crops, medicines, etc., thus creating many business and trade opportunities and boosting the economic growth of a country. But the question that comes into mind here is that whether this should be the idea behind conservation?

Utilitarian Versus Intrinsic Value of Biodiversity: As already discussed, the biodiversity conservation done for the sake of reaping certain material benefits comes under the narrowly utilitarian view, but a deep ecological idea would be to conserve Biodiversity for their intrinsic value. This means that whether any species is of any utility to humans or not, it is still being conserved and valued for its own sake and for the sake of ecosystem and its integrity as also for the individual rights that the species hold for survival and growth (Chan, Pringle, & Ranganathan, 2007).

There have been two contrasting views that have been found with regard to conservation of ecology and ecosystem: one being the 'shallow ecology' which gives more prominence to the immediate gains and utility from the ecology and environment in terms of material benefits like, wood, gum, fruits, vegetables, fodder and the like. The other and more philanthropic view is the 'deep ecology' that treats ecology and its resources not just as a means to an end of satisfying human hunger and greed, rather an end in themselves and possessing a complete right for existence (Naess, 1995). According to Arne Naess, a deep ecologist, who argues for the special role of deep ecology in sustaining life, said,

> For one, it rejects the monopoly of narrowly human and short-term argumentation patterns in favour of life-centred long-term arguments. It also rejects the human-in-environment metaphor in favour of a more realistic human-in-ecosystems and politics-in-ecosystems one. It generalizes most eco-political issues: from 'resources' to "resources for"; from life quality to "life quality for"; from "consumption" to "consumptions for" … where "for" is, we insert "not only humans, but other living beings." (Naess, 2005)

The idea that is being proposed here is in no way a shift from caring for humans to non-humans rather, what is required is a shift in perspective to view all life forms

Table 2.11 Ethical matrix for biodiversity conservation

Stakeholders	Consequentialist utilitarian	Deontological (duty and rights perspective)	Virtue	Care	Principles
Human society	Material resources and economic benefits	Community welfare		Care for all the species	No harm posed to any organism
Non-human mammals	Conservation Ecosystem stability	Animal rights	Intrinsic value of species		Equality
Environment/ ecology	Conservation of resources for future generations Building a more environmentally sustainable future	Intrinsic value of Biodiversity including plants, animals and other organisms and their right to live and right to being	Protection and conservation of each species of organisms on earth	Deep ecology principle: caring for all Biodiversity irrespective of the material benefits	

in totality and care for them alike (Naess, 1995). The philosophy of Mahatma Gandhi involving the practices of non-violence and political liberation also aligns with the deep ecology movement as it also extends beyond the diversity of humans to all living creatures, especially given in his principle of "*Advaita*" or non-duality (Slattery & Rapp, 2003, p. 210) (Table 2.11).

2.3 Conclusion

The nine issues discussed so far pertain to the potential areas for ethical inquiry within the curriculum of Biological Sciences, both at the high school and undergraduate levels. Each issue discussed is inherently ethical as depicted explicitly in the ethical matrices constructed for each of the issues detailing their reliance over the major ethical theories for decision-making. It is evident from the ethical matrices that sometimes two or more theoretical positions or arguments stand at the crossroads which can create an ethical dilemma. For example, regarding the issue of Biodiversity and conservation, justice to humans stand in conflict with Biodiversity preservation, especially with the formulation of conservation policies at the stake of local communities and their livelihood. A similar conflict of interests could be witnessed in the issue of 'Gene Therapy' where on one hand the technique could be useful in pharmacogenomics, then on the other hand the issue of 'genetically engineered babies' could trigger inequality in the society. Thus, the ethical matrix can in no way solve a particular ethical issue, rather provides a space for understanding the different conundrums raised by a particular technology or issue at hand. The decision-making with respect to any ethical issue or technology per se can be a very subjective opinion when it comes to individual decision-making owing to one's own ethical subjectivity, but needs to be supported with strong evidences if the decision has to become a policy directive, as, at the policy level, the decision on a particular technology is going to impact a larger section of the world population.

It is also pertinent to mention over here that the mention of so many ethical issues within the realm of Biotechnology does not mean that the research or Science is to be blamed. Rather, Science in itself is amoral as its only aim is quest for the 'Truth', which has been its aim since the establishment of Science as a separate discipline during the Renaissance movement (Muralidhar, 2019). It was only later on added by Francis Bacon and Rene Descartes that Science should also be useful to man for the development of comfort through control and exploitation of natural resources (Bernal, 1954). This defines the separate field of Science termed as Application-Oriented Science (or even Technology) where the real ethical issues arise.

References

Akbarsha, M. A., Zeeshan, M., & Meenakumari, K. J. (2010). Alternatives to animals in education, research and risk assessment: An overview with special reference to Indian context. In *Altex Proceedings 2, 1/13, Proceedings of Animal Alternatives in Teaching, Toxicity Testing and Medicine*. Retrieved from http://www.altex.ch/resources/rISC_001_Akbarsha2.pdf.

Bartholet, E. (1992). In-vitro fertilization: The construction of infertility and of parenting. In H. B. Holmes (Ed.), *Issues in reproductive technology I: An anthology* (pp. 253–260). New York: Garland.

Bernal, J. D. (1954). *Science in history*. London: Penguin Books.

Chalmers, A. F. (1999). *What is this thing called Science?* (3rd ed., pp. 19–26). Buckingham: Open University Press.

Chan, K. M. A., Pringle, R. M., & Ranganathan, J. (2007). When agendas collide: Human welfare and biological conservation. *Conservation Biology, 21*(1), 59–68.

Cooper, N. G. (1994). *The human genomic project: Deciphering the blueprint of heredity*. Mill Valley, CA: University Science Books.

Fletcher, J. C. (1983). Moral problems and ethical issues in prospective human gene therapy. *Virginia Law Review, 69*(3), 515–546.

Gannett, L. (2010). The human genome project. In *The stanford encyclopedia of philosophy*. Retrieved from http://plato.stanford.edu/archives/fall2010/entries/human-genome/.

Giridharan, N. V., Kumar, V., & Muthuswamy, V. (2000). *Use of animals in scientific research* (pp. 1–25). New Delhi.

Hill, C. M. (2002). Primate conservation and local communities: Ethical issues and debates. *American Anthropologist, 104*(4), 1184–1194.

ICMR (2004). Regulatory regimen for genetically modified foods: The way ahead. New Delhi.

Josefson, D. (2002). Scientists manage to manufacture Polio Virus. *British Medical Journal, 325*, 122.

Knight, A. (2009). Perceptions, knowledge and ethical concerns with GM foods and the GM process. *Journal of Public Understanding of Science, 18*, 177–188.

Kuhn, T. S. (1970). *The structure of scientific revolutions* (2nd ed.). USA: University of Chicago Press.

LaFollette, H., & Shanks, N. (1996). *Brute science: Dilemmas of animal experimentation*. London: Routledge.

Macklin, R. (1994). *Surrogates and other mothers: The debates over assisted reproduction*. Philadelphia: Temple University Press.

Mastroianni, Jr., L. (2003). Reproductive technologies. In *Encyclopedia of bioethics*. Macmillan Reference USA.

Mepham, B. (2008). *Bioethics: An introduction for the biosciences* (2nd ed.). New York: Oxford University Press Inc.

Muralidhar, K. (2019). Introduction. In K. Muralidhar, A. Ghosh, & A. K. Singhvi (Eds.), *Ethics in science education, research & governance* (pp. 1–18). New Delhi: Indian National Science Academy (INSA).

Naess, A. (1995). The Apron diagram. In A. Drengson & Y. Inoue (Eds.), *The deep ecology movement: An introductory anthology*. Berkeley, CA: North Atlantic Books.

Naess, A. (2005). Deep ecology and politics. In H. Glasser & A. Drengson (Eds.), *The selected works of Arne Naess* (pp. 189–226). The Netherlands: Springer.

Nelkin, D. (2002). Patenting genes and the public interest. *American Journal of Bioethics, 2*(3), 13–15.

Parascandola, M. (1998). Animal research. In *Encyclopedia of applied ethics*. Academic Press.

Pascalev, A. (2003). You are what you eat: Genetically modified foods, integrity and society. *Journal of Agricultural and Environmental Ethics, 16*, 583–594.

Regan, T. (2014). Animal rights and vivisection. In A. I. Cohen & C. H. Wellman (Eds.), *Contemporary debates in applied ethics* (2nd ed., pp. 95–108). UK: Wiley.

Rowland, R. (1992). *Living laboratories: Women and reproductive technologies*. Bloomington: Indiana University Press.

Scully, J. L., Haimes, E., Mitzkat, A., Porz, R., & Rehmann-Sutter, C. (2012). Donating embryos to stem cell research: The "Problem" of gratitude. *Bioethical Inquiry, 9,* 19–28.

Shiva, V. (1996). *Biopiracy: The plunder of nature and knowledge*. Boston: South End Press.

Slattery, P., & Rapp, D. (2003). *Ethics and the foundations of education: Teaching convictions in a postmodern world*. Boston: Pearson Education Inc.

Tumpey, T., Basler, C., Aguilar, P., Zeng, H., Solorzano, A., Swayne, D., et al. (2005). Characterisation of the reconstructed 1918 Spanish Influenza Pandemic Virus. *Science, 310,* 77–80.

Chapter 3
Ethical Issues in Science Curriculum—A Case of K-12 Biology Textbook

Abstract The chapter presents a critical content analysis of the K-12 Biology textbook with respect to the inclusion of Bioethical issues. Content analysis as a technique for analysis has been chosen to analyse the text. In simple terms, content analysis can be defined as the process of summarizing and reporting data, as well as the main concerns of that data and revealing hidden messages if any. The rationale behind writing this chapter is to highlight such areas and topics within the Biology Text where there could be some confluence of science, technology and Ethics as well as to critically analyse the textbook content for facilitating critical thinking and ethical understanding on the same.

Keywords Critical content analysis · K-12 biology textbook · Critical thinking · Ethical understanding

3.1 Introduction

Textbooks offer the most significant representations of the curriculum. However, the written texts within these textbooks can be subjected to multiple interpretations as noted by Apple (1991), such as in terms of 'dominant', 'negotiated' and 'oppositional' meaning-making. Therefore, how these text materials are being treated in the classroom is determinant of the level of engagement and critical interrogation of facts and issues given therein.

Text helps in giving a direction to the learning process, provided that it is not biased on any ground. It helps in the structuring of the teaching–learning process by way of guiding the teachers about the focal areas of concern, and sometimes suggesting even a probable effective pedagogy. At the level of school education, textbooks are considered to be the most essential and infallible texts which are followed both by the teachers and students religiously with little opportunity for the teachers to put forth their judgment about the text and the content given therein. They are constantly being upgraded and modified by the higher authorities such as the policymakers and curriculum developers to meet the demands of the globalized

A. Saxena, *Ethics in Science*, https://doi.org/10.1007/978-981-32-9009-9_3

world. In the light of the National Curriculum Framework (NCF-2005), some major modifications have been proposed for the textual content of Science textbooks, by making them more learner-centric, including more activities, puzzles, linking to daily life examples and analogies, as well as building of concepts in a spiral fashion rather than aggregation of facts, definitions and figures. NCF 2005 refers to 'ethical validity' as one of the important parameters that Science curriculum should adhere to. According to it, the Science curriculum should develop and promote the values of honesty, objectivity, cooperation, freedom from fear and prejudice and concern for life and preservation of the environment. Therefore, keeping in mind the above backdrop, the present chapter aims at analysing the K-12 Biology textbook from an ethical lens.

3.2 Content Analysis-Method and Approach

The objective of the present chapter is to critically analyse the content of class XII NCERT Biology textbook for the inclusion and appropriate addressal of Ethical Issues; therefore, content analysis as a technique for analysis was adopted. In simple terms, content analysis can be defined as the process of summarizing and reporting data, as well as the main concerns of that data and revealing hidden messages if any (Cohen, Manion, & Morrison, 2007). Content analysis as a research technique could be of great help in interpreting and generating the intended or implicit meaning from the text. However, there could be many ways of interpreting and analysing a given text that makes it all the more subjective and reader-dependent (Krippendorp, 2004, pp. 22–4). This could be one of the limitations of content analysis that makes it less objective as a research method. In order to overcome this lack, certain quantitative attributes have been added to this technique, such as coding and statistical means of generating frequency tables for coded responses or themes, counting the number of words or repetition and occurrence of a particular theme or phrases, etc. It has now been acknowledged that the highest quality content-analytic studies involve both quantitative and qualitative analyses of texts (Weber, 1990, p. 10).

The building up of an ethical understanding has to begin in the formative years of a person's education. However, a mature understanding of these controversial and ethical issues requires a certain cognitive level that is well achieved beyond the formal operations stage (*Piaget's theory of stage wise cognitive development*, pp. 71–73, Woolfolk, 2004) and post-conventional moral reasoning state or 'Universal ethical principle orientation' that marks the realization of individual conscience and understanding of abstract concepts such as justice, human dignity, equality, etc. (*Kohlberg's stages of moral development*, pp. 114–115, Woolfolk, 2004). The context also plays a pivotal role, as when the topics related to such issues are being taught, then it becomes easier for a learner to comprehend not just the semantic meaning of the terms but venture deep into the relational significance such as impact of a particular technology on the society at large and the decision-making on the appropriate usage of that technology by weighing the pros and cons. It is over here that a textbook can offer some amount of integration and flexibility within the curriculum. This is to say, not restricting the content of the textbook to specific subject boundaries rather moving beyond it so as to include the sociopolitical, economic and ethical perspectives. This kind of a classification is of an open-type and provides scope for interaction between different subject contents due to weak boundaries ('Concept of Classification and Framing of Content', Bernstein, 1971).

Therefore, the rationale behind writing the present chapter is to highlight such areas in the Biology Textbook where there is possibility of integration between different subjects and disciplines such as, science, technology and Ethics so as to think beyond the literal meaning of the text and gain a deeper understanding of the issues embedded therein. This would also enable the learners to develop their critical thinking and ethical understanding on these socio-scientific issues.

3.3 Methodology

The methodology proposed for analysis is qualitative in approach and descriptive in character. This has been done in the following parts:

Part-1(a) Review of Relevant Documents for deriving the basis of the study.
Part-1(b) A review of the previous editions of K-12 Biology textbooks (from 1960s to 2002).
Part-2 Interviews of members from textbook/s writing committee.
Part-3 Critical Analysis of K-12 Biology textbook with respect to the inclusion of Ethical Issues.
Part-3 Compilation and comparison of findings both from primary and secondary sources.
Part-4 Suggestions and implications for textbook writing for future.

3.3.1 A Review of Documents

The following documents have been reviewed for their position on the inclusion of ethical issues in Science:

- National Curriculum Framework (NCF)-2005;
- Position Paper on Curriculum, Syllabus and Textbooks;
- Position Paper on Teaching of Science.

NCF-2005

A brief mention has been made about inculcating an ethical understanding amongst the students by the way of science curriculum and textbooks in an integrated manner,

> the ethical understanding needs to be interwoven into subjects and activities. (p. 29, NCF-2005)

Even one of the basic criteria for the validity of the science curriculum involves the component of "Ethical Validity" and the document defines it as

> that the curriculum promote the values of honesty, objectivity, cooperation, and freedom from fear and prejudice, and inculcate in the learner a concern for life and preservation of the environment. (p. 48, NCF-2005)

However, the document does not seem to propose any specific route through which this integration can be made possible. It also seems to project a limited understanding of the ethical component in the curriculum, which can be traced up to ethical decision-making, identification as well as solution to various ethical issues plaguing the world at large. The principles of ethical validity outlined in the document are mainly moral stances and pertain to general normative and behavioural objectives.

National Focus Group on Curriculum, Syllabus and Textbooks (*Position Paper National Focus Group on Curriculum, syllabus and textbooks*, 2006)

The document focuses on the ideal design and development of curriculum, syllabi and textbooks for all the subjects. The document propagates that science textbooks need to be universal in approach throughout the country and should be devoid of any kind of biases. There is a mere mention about the development of a general ethical understanding amongst students, but how that needs to be integrated into the curriculum is missing. This has to be integrated and ingrained in the textbooks so as to enable the students in ethical understanding and decision-making.

One important point that the documents mention is the introduction of concepts at the level appropriate to the learners' cognitive age and should neither be introduced too early nor too late.

Position Paper on Teaching of Science (*Position Paper National Focus Group on "Teaching of Science,"* 2006)

One of the chief aims of Science Education as dictated by the document is to imbibe values of honesty, integrity, cooperation, concern for life and preservation of the environment. This aim clearly reflects an ethical understanding and way of life to be imbibed by the students. Although the document is affirmative towards the inculcation of the issues at the interface of science, technology and society at the classroom discussion level, it forbids their formal assessment.

Another important point that the document raises is the fact that teachers usually do not refer to these National-level documents for classroom transactions and rely solely upon the textbooks; therefore, it is important for the textbook writers to seek for the translation of curricular goals in terms of textual content.

Discussion

Reviewing the above documents, it is evident that ethical issues as well as an ethical understanding do not receive an explicit mention in the curricular aims. Owing to the recent developments in science and technology, these issues have acquired serious implications in the lives of humans. This has resulted in an emergent need to generate an adequate understanding of the same so as to make informed decisions in daily life, only then can science and its practices acquire a liberating potential. Although the documents do mention about the coverage of issues at the interface of science, technology and society at the level of classroom transactions by way of discussions and debates, their inclusion in assessment and evaluation scheme is still being eluded. Till they are made the part and parcel of the formal curriculum and acquire a space in the textbooks they will not be taken seriously. As we know ours is a multicultural country, therefore it is important to relate the content of science textbooks with the prevailing context of that particular region, for which we require multiple textbooks. This has been explicitly stated in the position paper on science teaching. In this way, knowledge can become true for the learner and true to nature.

3.3.2 Review of the Biology Textbooks (Period: 1960 to the Present)

The present section deals with a cursory review of the older versions of the Biology textbook for K-12 in the context of inclusion of ethical issues. While analysing

Table 3.1 Analysis of previous biology textbooks in the light of major policy documents

Time period	Overview of the Biology Textbook in terms of inclusion of ethical issues	Major policy documents	Recommendations
Late1950s–1970	Major thrust is upon the origin of Science, practice and application of the method of Science, fostering scientific attitude amongst students, Science being viewed in a positivistic sense and as value-free. Technology, wherever mentioned, is presented as a fruit of applied sciences and a boon for society in terms of solving many problems such as population explosion, food shortage, improving livestock for better yield, treatment of many deadly diseases by the discovery of antibiotics.	Science Policy Resolution—1958 Kothari Commission (1964–66) Second Five-year Plan (1956-61); Third Five-year Plan (1961-66)	Intense cultivation of Science and scientific research at a large scale to meet the country's requirements; encourage, and initiate, with all possible speed, programmes for the training of scientific and technical personnel, on a scale adequate to fulfil the country's needs in science and education, agriculture and industry, and defence (Dhar and Saha, 2014) Science and technology as one of the important features of modernization, but not at the stake of killing social, moral and spiritual values Science is a dynamic power and has a vitalizing influence. It has improved standards of living and provided a basis for the welfare state. Technology as a key to national prosperity. Science policy aims at creating opportunities for research and creativity for the benefit of mankind. Establishment of new technological Universities and research centres, such as IITs to promote scientific research and development.

(continued)

Table 3.1 (continued)

Time period	Overview of the Biology Textbook in terms of inclusion of ethical issues	Major policy documents	Recommendations
1971–1980	Focal areas emerging during this period are Genetics, its role in heredity and how the techniques of Genetics can be used in the improvement of traits in plants and animals alike. Thus, again a positive picture of technology is being portrayed	Fifth Five Year Plan (1974-79); Indian Patent Act (1970)	A sectoral approach was adopted to ensure interaction between research agencies in order to facilitate technology transfer.
1981–2000	Technology has been given increasingly more space in the textbooks by way of advancement in medical technology, and agricultural Biotechnology	National Policy on Education—1986 Technology Policy Statement—1983; Sixth Five-year Plan (1980-85); Seventh Five-year Plan (1985-90)	- A greater stress on the role of research institutions fostering development in the sphere of science and technology - Technological self-reliance and resistance to technological imposition from outside Technology advancement viewed as a social and political tool; technology assessment, development, acquisition, absorption, utilisation and diffusion
2000–Present	Some of the ethical issues start featuring in the textbooks such as those under the subheading, "Misuse of Biology" which discusses Amniocentesis being misused for sex-determination and leading to female foeticide, also creation of bioweapons leading to bioterrorism (NCERT, Biology textbook, 2002). Some other issues of ethical concern have also been incorporated in the later versions such as Biopiracy and Bioethics (NCERT, Biology for Class XII, 2003)	Science and Technology Policy (2003), NCF-2005 Position Paper on Teaching of Science, Eleventh Five-Year Plan (2007–2012), Vol. II Science, technology and innovation policy (STI)-2013	Science and technology are becoming more complex by not remaining an individualistic enterprise but a big corporate effort. Scientific knowledge as a source of economic and power being recognized Besides, for human and environment concerns, an added focus is now on creating wealth and global competitiveness 'Ethical Validity' towards the inculcation of basic human values, such as truth, honesty, etc. Stresses upon the development of an integrated science and technology plan, fostering the sanctioning of funds for research institutions, and encouraging research in science and technology Science, technology and innovation as major drivers for national development globally. Recognition of Indian society as a major stakeholder in the science and technology research for development. Aims at bridging the gaps between the STI system and socio-economic sectors by developing a symbiosis between the two

the different editions and versions of the textbook, it can be surmised that the content of the textbook is in consonance with the major reports and policy documents of that time. This clearly depicts the role of authority (which in the present case is the Ministry responsible for giving directions on textbook development at different stages) in determining the curriculum of the country. Therefore, an attempt has been made to evaluate the textbook content in the light of these reports and Government policies which would facilitate a better conceptualization of the larger debate for the inclusion of ethical issues in the Biology textbook. This has been done in a timeline fashion by drawing comparison between the major content areas covered in the past editions of the Biology textbook and the major policies during that time along with their recommendations. (refer to Table 3.1).

The textbooks reviewed clearly indicate a gradual shift in terms of viewing science and technology from core science in the 60s towards the applied science and technology in the 70s and 80s. This change can be construed as the impact of the industrial revolution on third world countries which is very well evident in the Government policies during this period. It can be concluded that textbooks of the 1950s and 1960s presented a 'neutral' view of Science, detached from moral or ethical constraints (Lynch & Paul, 1985). Modern Science (late twentieth and early twenty-first century) is presented within a value-laden framework and attempts to raise arguments against the misuse of technology. But, how far have they actually succeeded in generating an ethical understanding on such controversial issues still needs to be assessed.

3.3.3 Opinion Generated by the Experts

Some members from the textbook-writing committee were contacted and interviewed for their opinion about the inclusion of ethical issues in the present K-12 Biology textbook.

Interview 1

The first expert who was interviewed presented a highly positive image of the textbook as well as its content by saying,

> The textbook is being written in such a manner that it reduces load on the learners by, material is learner friendly, interactive, information load reduced, core contextualized.

According to the expert's view, the textbook caters very well to the various kinds of validities that the NCF-2005 talks about, including content validity, context validity, historical validity, cognitive validity, process validity and ethical validity. When specifically asked as to how this ethical validity is achieved in the textual content, which is of major concern here, the expert, replied,

> Wherever we could find, we have addressed the ethical validity by taking care of the ethical issues in the K-12 Biology textbook, but one has to view the things in totality such as K-11 and K-12 syllabi together as there we have tried to deal with the physiology of all living beings and various organs and organ systems.

The above description of ethical validity given by the expert appears to be highly ambiguous and does not explain as to how this ethical validity has been taken care of in the present version of the textbook.

When asked about the role of textbooks in generating an ethical understanding, the expert put the onus on the teacher by saying,

> Textbook should be taken as a tool, the perception of teachers has to change, it is not like you administer a drug and you get a response.

As we know, in the Indian scenario, teachers do not have a say in the making and amendment of the curriculum. Therefore, till these issues do not reach the textbook level, they will not be taken up seriously either by the teachers or by students. Although efforts of teachers in this regard are indispensable, along with that we need to provide some supplementary text to both the teachers and students focusing upon these Bioethical issues.

Interview 2

One of the experts mentioned about NCF-2005 that talks about honesty, preservation of environment and conservation of resources. Although such a statement gives a broader purview of ethical validity, it does not adequately address how these are being taken care of in the textbook, or certain instances which can confirm their incorporation in the present K-12 Biology textbook. The expert argued,

> The issues have been discussed but may be some people want it in a different way, especially people who study social sciences.

The above statement depicts the shifting share of responsibilities, and kind of guarding the subject and creation of boundaries between science and social sciences which is unacceptable as of now when we talk more of integration of concepts and an interdisciplinary approach even in sciences.

Another point from the expert's side that highlighted her apathy towards the treatment given to these issues in the textbook was,

> At K-12 standard, the students are into a subject now and the general education is over.

Thus, ethical issues still do not receive the kind of treatment that they deserve, they are rather segregated from the core sciences and made a part of general science education which seldom generate the kind of reasoning and sensitivity required in building an authentic ethical understanding and decision-making in this regard.

The expert seemed to cohere with the point of view put forth by the earlier expert when she said,

> a textbook is not sufficient to generate an adequate understanding about these issues, but is the task of a teacher and other teacher education programmes or teacher refresher courses to guide them in this direction.

Thus, limiting the role of a textbook to inform and educate just about the major scientific concepts, theories and laws and refraining the textbook from addressing the ethical issues would offer a biased perspective about Science and Technology.

When asked about the ways by which the present textbooks can give some more space to these issues, the expert replied,

> Textbooks have limited space and everything cannot be included in them, also effort is being made to keep the language of the textbook non-controversial so that it does not hurt the sentiments of the people. Therefore, the issues that appear to be controversial at the very first instance are deleted, for instance, Animal Experimentation.

It is quite agreeable that textbooks have a limited space, but raising meaningful concern about these issues does not require page-length words; even a single statement can generate enough discussion and debate in the classroom if addressed appropriately. The problem lies somewhere in our approach and intent rather than on structures and norms. An attempt to keep the language and intent of the text non-controversial also keeps these issues at bay, as was being emphasized by the expert,

> Vignettes could have been quoted but due to certain sociopolitical influences these cannot appear in the textbook, as they tend to magnify a particular region, religion, cast or class.

The expert is well aware of the ethical concerns and has a wider background for ethics which appears right when she said,

> When we talk about ethical validity, why we are just focusing on animals and not plants? People involved in ethics are anthropocentric. Every issue should be taken from an animal's point of view as well as from a plant's point of view. We allow anything happen to plants, such as creation of hybrids.

One of the ethical issues that the expert herself mentioned related to the language of the text was on one of the sexually transmitted diseases.

> A blunt statement made in the textbook of class XII that says that people are dying of AIDS, and leaves a very little hope for them; instead the issue should have been dealt more sensitively.

Thus, the expert is aware of some flaws that the textbook writers have made and wishes to modify them as well but is still not completely convinced with the idea of including ethical issues in the textbook.

Interview 3

It seems that there is a gap in the understanding about the working in a school and the role of textbook both for the students and the teachers, which is based on the statement,

> Addressal of Ethical issues is not the objective of textbook, and wherever it's been done, it is an error.

This clearly indicates that the expert tries to present an objective and value-neutral picture of science to the students and forbids from the inclusion of any emotive element.

There is, however, an agreement over the fact that science and technology and their application in human life raises some ethical concerns, especially when the expert said,

> Some of the recent developments in Science and technology are 'potentially unethical', due to misinformed or misplaced emotion.

But, the expert disavows their inclusion in the textbook due to the fact that it is not science in itself which is ethical or unethical rather its misused applications; here, the expert even quotes the example of an atom bomb which is the most dangerous invention using scientific principles. The following statements from the expert's side make his stance come out more clearly,

> Science in itself is 'amoral', but wherever a human element and his survival is at stake, then only the question of 'ethics' arises, that is to judge whether a particular act is right or wrong.

A point is also made that ethical issues should be made a part of pre-service teacher education curriculum rather than bringing them up in the National-level textbooks.

Discussion

The recent curriculum modules such as STS Curriculum and Nuffield Science Curriculum have courses centred around these ethical issues for K-12 students, and have proven statistically that students' outcome in terms of grades, attention span, engagement and concept development is higher in these courses than the conventional modules where these elements are missing. By saying that wherever there happens to be a brief mention about ethical issues is by mistake, is a clear-cut indication that ethical issues do not receive much of importance and space in the curriculum and have been obliterated intentionally. Thus, opinion generated by the experts is mostly in justification of the present textbook of Biology at the K-12 level and the treatment given to ethical issues therein. The onus is often being transferred on to the teacher education programmes, refresher courses for the teachers or the pre-service teacher education curriculum rather than the mainstream science curriculum. This offers a limited view, as just by introducing these issues at the pre-service level cannot develop an authentic ethical understanding among the learners at school level. As, there are large number of teacher education programmes being run by different teacher education institutions that might not be offering the same quality and intent of the course. Besides, developing ethical understanding in itself is a long-term endeavour and requires continuous and rigorous engagement with these issues at different levels. Also, in pre-service education, the main focus is on pedagogy and less on content development, and making ethical issues a part of teacher education syllabi is sidelining the content validity of these issues.

3.4 Critical Content Analysis of K-12 Biology Textbook

The present chapter is delimited to the analysis and discussion of only those areas where a confluence of science and ethics can be explicitly seen and visualized and that pertain to the realm of Socio-Scientific Issues (SSI). The first phase of this study includes the identification of such issues in the textbook. In order to do this, a list of categories of Ethical issues was prepared (Appendix A) so as to give a detailed account of the range of ethical issues being covered under the ambit of Biological Sciences and applied Biotechnology. Thereafter, an ethical issue from K-12 Biology textbook is picked up and analysed with regard to the ethical content being addressed in it. A rubric has been designed in order to state the chief attributes of an issue with respect to its ethical dimensions (as shown in Appendixes A and B).

3.4.1 Ethical Issue: Amniocentesis and Sex-Determination

Type of ethical issue	Medical technology and its misuse
Category no.	2.1
Broad area	Reproductive health
Length of issue covered in textbook	32 words
Reference to any ethical theory	None
Arguments raised	One
Kind of understanding fostered	Factual

The above description clearly depicts that the issue of sex-determination via amniocentesis is being taken up in a superficial manner in the textbook, by just putting facts and that too with missing details on the scientific technique or procedure. The critical arguments raised by the medical professionals, feminist activists and certain social groups for the protection of human rights also appear to be missing. The issue fails to acquire an ethical validity in the textbook as it does not generate any ethical understanding in the minds of the students. The precise content from the textbook says,

> Statutory ban on amniocentesis (a foetal sex determination test based on the chromosomal pattern in the amniotic fluid surrounding the developing embryo) for sex determination to check increasing female foeticides, massive child immunization, etc. (NCERT, Biology, 2006).

Upon carefully scrutinizing and deconstructing the above text, it is evident that the technique of amniocentesis is being misused for foetal sex-determination leading to increasing number of female foeticides. However, the social causes of the skewed sex ratio in our country are missing from the text. Hence, the text is not able to generate adequate sensitivity amongst the students. Also, such an argument generates a one-sided view of the problem, debarring the use of technology completely while not considering the positive aspect of it, which is the determination of certain foetal genetic disorders.

Ethical Basis: The ethical grounding for such an issue could be the principle of Equality of gender as per which both male and female should receive equal human rights. The right to life is one of those inalienable rights irrespective of class, creed, religion, gender or region which should be granted to everyone. The ethics lies in using this technique just for diagnosis and not for sex-determination. Banning the technique altogether has not been able to improve the skewed sex ratio, rather an ethical understanding of wisely using the technique can. The responsibility of the medical professionals is to follow a code of ethics in their profession by not divulging the sex of the foetus while carrying out this kind of test.

3.4.2 Ethical Issue: Population Explosion and Birth Control

Type of ethical issue	Medical
Category no.	2.2
Broad area	Reproductive health
Length of issue covered in textbook	$3^{1/2}$ pages
Reference to any ethical theory	None
Arguments raised	Nil
Kind of understanding fostered	Conceptual

The above issue deals with a sensitive area of sexual reproduction and birth control by using contraceptive devices. The main argument given here that goes in favour of the usage of these contraceptive devices is to control the geometrical rate of explosion of population.

> According to the 2001 census report, the population growth rate was still around 1.7 percent, i.e., 17/1000/year, a rate at which our population could double in 33 years. Such an alarming growth rate could lead to an absolute scarcity of even the basic requirements, i.e., food, shelter and clothing, in spite of significant progress made in those areas.

The argument proposed is a valid and logical one in terms of the limited resources, sustainability, and individual growth and development prospects. The textbook briefly describes the different methods of contraception including natural, barrier, intrauterine devices, pills and surgical means. The textbook professes that, “an ideal contraceptive is that which is user-friendly, easily available, effective and reversible with no side effects”. However, the content within this topic does not address the ethical issues associated with the use and misuse of these contraceptives. Thus, it tends to generate only a conceptual and need-based understanding and not an ethical one.

Ethical Basis: The ethical stance involved here is that of interfering with nature’s processes and also killing off the human embryos that had the potential of developing into full-fledged human beings. Therefore, it has been argued by many ethicists that only such methods should be used that prevent the formation of the embryo itself, such as barrier methods or natural methods (abstinence), rather than those that prevent conception at the implantation level (these will include intrauterine devices, etc.). Thus, the end result of all the methods of contraception is the same, i.e. avoidance of conception, but the means chosen are different and hence it follows a deontological ethics to resolve the issue. Another Scientific argument in favour of these contraceptive devices, especially IUDs (or intrauterine devices), however, also states that IUDs are non-abortifacient, rather they interfere with the process of fertilization, by immobilizing the sperms or creating unfavourable conditions for the proper functioning of sperms in the female genital tract. The Roman Catholic Church forbids the use of any means of contraception as it regards children to be God's most precious gifts, and any kind of sexual activity without the intention of bearing a child is deemed unethical. However, this notion in contemporary times is being contested as the quality of life has become more important along with the family's overall health. Therefore, the practice of family planning came into being providing grounding for the use of contraceptives for family planning. On the other hand, the responsibility of communicating the pros and cons of each kind of contraceptive lies with the health professionals. This is especially in the case of emergency contraceptives as many women using oral contraceptives suffered some side effects such as anaemia, goitre, kidney and heart diseases (Tadiar & Robinson, 1996).

3.4.3 Ethical Issue: MTP or Abortion

Type of ethical issue	Medical
Category no.	2.3
Broad area	Reproductive health
Length of issue covered in textbook	½ page
Reference to any ethical theory	None
Arguments raised	Two
Kind of understanding fostered	Ethical dilemma

The Abortion or MTP as stated in the textbook is usually carried out in case of unwanted pregnancies or where carrying on with it could lead to harmful consequences either for the mother or foetus or both. The arguments raised in the text include the following.

> MTP has a significant role in decreasing the population though it is not meant for that purpose. (p. 62)
>
> Another dangerous trend is the misuse of amniocentesis to determine the sex of the unborn child. Frequently, if the fetus is found to be female, it is followed by MTP-this is totally against what is legal. (p. 62)
>
> MTPs are also essential in certain cases where continuation of pregnancy could be harmful or even fatal either to the mother, or to the fetus, or both. (p. 62)

The arguments clearly posit that MTP is being extensively used for killing female foetuses in the desire for a male child although that is not the primary objective of this technique. In certain situations, the government also gives the right to the women to go for abortion, such as certain medical complications in case the mother is carrying an abnormal foetus or pregnancy resulting from a sexual assault. However, the ethical question implicit in this argument is that whether MTP is a desirable medical technology? This is because the implications are for both the diagnostic and deterministic fates of the foetus.

Ethical Basis: The issue derives its ethical connotation from the human rights perspective according to which it is unfair rather heinous to kill the foetus that has the potential to develop into a full-fledged human being. However, in cases where the existence of foetus or the mother is dubious, such as any kind of deformity in the foetus and teenage pregnancy, then the option only available is abortion. In cases of sexual harassment also, the Government gives complete liberty to the mother-to-be to exercise her reproductive right and go for abortion. Much of the ethical debates related to abortion centre are around the fact that whether the foetus is a person or not, as the concept of personhood only relates foetus to the most inalienable right to live (Hinman, 2009). Another concept is that of sentience which regards the foetus as a completely sentient being capable of experiencing pleasure and pain that according to some neurophysiologists is only attained during the second trimester of pregnancy.[1] The argument of potentiality as stated by some philosophers is also debatable. Some state that the Zygote formed immediately after fertilization having a complete set of chromosomes can be regarded as a potential human being, while others regard only the ex vivo condition of the foetus, which means outside the mother's uterus as having the actual potential for a sufficiently surviving human being and a potential candidate for exercising and using his/her rights (Lotz, 2005).

The autonomy and feminist rights perspective bestows certain ethical, moral and human rights for women that grant her the freedom to decide upon her fate, such as

- Right to Privacy,
- Right to the Ownership of One's Own Body,
- Right to Equal Treatment,
- Right to Self-Determination.

These rights are in favour of woman and place the onus of taking a decision about the sustenance of the foetus on her without anybody's forceful intervention.[2] Thus, the woman has every right to take a decision and control what happens in and to her body. Hence, it cannot be said always that Abortion is an unethical act, rather one should always weigh the pros and cons and only then take an ethically appropriate decision (Guided by the Utilitarian Perspective).

[1]Refer to Mary Anne Warren, 'Abortion', in Peter Singer (ed.) 1993, *A Companion to Ethics*, (Cambridge, MA: Blackwell): pp. 308–9.

[2]For further readings refer to Schwartz, S. (1990). The Moral Question of Abortion. Loyola University Press. And Bayles, M. (April, 1990). Feminism and Abortion. *Atlantic Monthly*.

Thus, the text does not address these ethical issues associated with abortion or MTP rather it tends to present it as a means or treatment to get rid of unwanted pregnancy although the health and medical aspects of it have been described.

3.4.4 Ethical Issue: In Vitro Fertilization (IVF) and Assisted Reproductive Techniques (ARTs)

Type of ethical issue	Medical
Category no.	2.4
Broad area	Reproductive health
Length of issue covered in textbook	1 page
Reference to any ethical theory	None
Arguments raised	One
Kind of understanding fostered	Conceptual

The textbook introduces the issue as a repair mechanism for infertility by offering medical technology that can overcome this problem.

> The couples could be assisted to have children through certain special techniques commonly known as assisted reproductive technologies (ART). (p. 64)

The different kinds of medical interventions in this regard have been explained as embryo transfer technique, Zygote intra-fallopian transfer (ZIFT), Gamete intra-fallopian transfer (GIFT), Artificial Insemination (AI), etc. Although the technicalities of all these techniques have been explained, what the textbook lacks is the whole set of ethical issues associated with these ARTs. The only argument that the textbook refers to is pertaining to social and economic inequality amongst people in a country like India, in terms of those who cannot afford or have access to such treatment.

> These facilities are presently available only in very few centres in the country. Obviously their benefit is affordable to only a limited number of people. Emotional, religious and social factors are also deterrents in the adoption of these methods. (p. 64)

Thus, the text alludes to the many emotional, religious and social factors that pose a hindrance in the complete adoption of the technology but does not discuss them in detail.

Ethical Basis: Although the technology of IVF is a medical innovation and an example of cutting-edge creativity, every good thing comes with a cost, so is this technology. The major issue lurking here is the doomed fate of many fertilized eggs or embryos which are either wasted or deep-freezed for further usage. In the United Kingdom under the Human Fertilization and Embryology Act (1990), human embryos can be kept under freezed condition only up till 5 years, after which they must be either used up or destroyed. This destruction of human embryos is being regarded as an extremely immoral act, as, according to some people, human embryos possess an equal moral status as any foetus, child or adult and therefore should be protected (Levinson & Reiss, 2003). Another ethical issue in in vitro fertilization is its use even by the fertile, such as single-parent family, delayed motherhood or postmenopausal motherhood, same-gender couples, etc.[3] Such kind of practices generate an ethical dilemma, such as what would happen to the children born to such parents, will they have a normal childhood, will their capacities be fully nurtured and developed? Here the principle of Utilitarianism prevails, which favours the act that leads to wellness and happiness to all. And an act such as in vitro fertilization and further development of embryo till the term is a highly responsible act and hence the consequences should be well thought of. Only after weighing the pros and cons, the technique should be practiced.

3.4.5 Ethical Issue: GM Crops, Gene Patents and Biopiracy

Type of ethical issue	Socio-economic
Category no.	3.1 and 3.3
Broad area	Biotechnology
Length of issue covered in textbook	2 pages
Reference to any ethical theory	Moral/virtue theory
Arguments raised	Ethical, socio-economic, legal arguments are being raised
Kind of understanding fostered	Ethical dilemma

The textbook presents the various ethical arguments related to genetic modification in a separate section titled "Ethical Issues", where the major focus is on GM Crops. The issue of morality or ethics is defined here in terms of the unforeseen repercussions of genetically modified organisms on the environment which can be quoted from the text itself.

> Genetic modification of organisms can have unpredictable results when such organisms are introduced into the ecosystem. (p. 213)

This has been the rationale behind the setting up of various organizations such as GEAC (Genetic Engineering Accreditation Committee) that ensure certain safety standards before marketing a particular genetically modified product. But what exactly are these safety standards is not being taken up in the present textbook. The textbook also quotes certain evidence of gene patenting and Biopiracy by Foreign multinationals for earning a profit, such as in the case of Indian Basmati rice and Indian traditional herbal medicines, e.g. turmeric and neem.

> In 1997, an American Company got patent rights on Basmati rice through the US Patent and Trademark Office. This allowed the company to sell a new variety of Basmati in the US, and abroad. This 'new' variety of Basmati had actually been derived from Indian farmer's varieties. (p. 214)

[3]"The Back-Up Plan" in Hindustan Times: June 11, 2012, New Delhi.

Thus, cheating on the indigenous farmers and people of the native region from where a particular crop or resource originated. Here, the ethical issue which could be raised is the ethics of the virtue of trust, justice and honesty that has not been appropriately addressed in the textbook.

The legal issue being raised is in terms of availability and formulation of certain patent rights to curb this malpractice on the part of developed countries in order to protect the rights and freedom of developing nations, thus initiating an ethico-legal debate in the following manner:

> There has been growing realization of the injustice, inadequate compensation and benefit sharing between developed and developing countries. Therefore, some nations are developing laws to prevent such unauthorized exploitation of their bio-resources and traditional knowledge. (p. 214)

Thus, the present textual content provides a food for thought in thinking about this issue of GM crops from an ethical perspective.

Ethical Basis: In order to make ethically appropriate decisions, some more ethical concerns need to be incorporated, and in an exemplified manner linking them to students' daily life. The following are the chief issues of concern pertaining to GM crops that have not been touched upon in the present textbook.

Food Quality and Nutrition: Genetic modification of plants may result in alteration in the nutritional profile of the plant product which can also result in altering the nutritional status of the consumer. For instance, Low glutelin genetically modified rice which is created using an anti-sense technique, signifies improvements in rice storage proteins for commercial sake brewing. The decrease in glutelin levels was however associated with an unintended increase in levels of prolamins. This illustrated that a targeted change in the level of a specific protein can result in other proteins being affected. The change in prolamin levels did not affect the industrial application but could affect nutritional quality and allergenic potential if the rice were used as a food. Other examples were given regarding soybean and rice showing that genetically modified soy with increased lysine showed an unexpected decrease in oil content, and the genetically modified "golden rice" designed to express betacarotene unexpectedly accumulated xanthophylls. This can result in a nutrient imbalance in the body as well as impact the overall dietary intake (WHO, 2000). Currently developed plants with improved nutritive value include GM rice with enriched vitamin A and GM soybean and rapeseed with modified fatty acid.

Food Safety: The inserted gene, or even the insertion process itself, may re-engineer the biology of the plant and generate poisons or toxins.[4] The case of GM potatoes experiencing *Galanthus nivalis* lectin gene for insecticidal properties is an example of the potential of GM foods to cause toxicity. For example, in a group of rats fed

[4]Refer to Purkayastha, P. and Rath, S. (2010, May 15). Bt Brinjal: Need to Refocus the Debate. *Economic and Political Weekly XLV (20), 42–48.*

with GM potato caused damage to their immune systems and stunted growth, and the experiment had generated considerable controversy (ICMR, 2004).

Gene Pollution and Loss of Biodiversity: The contamination of non-GM varieties of plants through pollen drift can cause loss of biodiversity. This was the reason behind the disapproval of GM Corn variety for commercialization in Mexico, as the native corn varieties might get contaminated by the foreign genes.[5]

Antibiotic Resistance—Potential for Gene Transfer: Sometimes, it has also been reported that GM food (particularly Bt Brinjal) leads to the generation of an antibiotic-resistant protein leading to alteration in blood chemistry including blood clotting time (prothrombin), total bilirubin (liver health) and alkaline phosphate in goats and rabbits.[6]

Emergence of Superweeds: As per some recent reports on GM field studies, the herbicide-resistant gene that is being genetically transferred to the GM crop can sometimes cross the species barrier and get integrated with the genome of some wild relatives of GM crops, which then become resistant to the effect of pesticides or herbicides.

Mixing of Genes—"Breach of Religious Faith": Many religions have explicit dietary prohibitions against certain foods or consuming particular foods (Pascalev, 2003, taken from Knight, 2009). Consensus conferences in Australia and the United Kingdom highlight lay concerns about mixing human and animal DNA with plants, being seen as tantamount to cannibalism.

Agricultural Knowledge Dissonance Leading to an Upsurge in Farmer Suicides: This is not to say that GM seeds are the sole cause of farmer's suicide, but one of them, and can be explained on account of agricultural knowledge dissonance. This is increasingly leading towards rejection of indigenous methods of production, and following the suite of developed nations by adopting advanced agri-based technology leading to farmer distress.

All the aforesaid arguments portray a different picture altogether about GM crops and raise ethical issues related to safety, environmental degradation, breach of faith and values, as well as loss of indigenous skill and labour. The utilitarian view will include weighing the pros and cons of a particular technology, and then making a decision which is in the benefit of all. This in no way means a majority opinion rather incorporating and collaborating with the viewpoint of the minority communities as well. The deontological ethics implies adherence to the laid down norms of conduct of the society and abiding by the safety principles while conducting experiments with GM crops.

[5]Refer to "India says No to Bt-Brinjal", retrieved from http://www.bhoomimatha.com/india-says-no-to-bt-brinjal/.

[6]Rejection of indigenous methods of production, and following the suite of developed nations by adopting advanced agri-based technology, whether they comply with Indian climatology and topography or not. See Vaisavi, A. R. (2004). Suicides and the Making of India's Agrarian Distress. National Institute of Advanced Studies, IISc Campus, Bangalore, India.

3.4.6 Ethical Issue: Genetically Modified Organisms (GMOs) and Transgenic Animals

Type of ethical issue	Therapeutic
Category no.	3.2
Broad area	Biotechnology
Length of issue covered in textbook	1 page
Reference to any ethical theory	None
Arguments raised	Positive arguments in the favour of GMOs
Kind of understanding fostered	Lopsided and unbalanced

The current issue is titled 'Transgenic Animals' and describes the genetic mechanisms involved in their creation by way of genetic or DNA modification by insertion of foreign genes in organisms giving rise to special characteristics. But the arguments being posed in the text are more towards the beneficial end, such as studying the effect of a particular gene insertion, pathology of a particular disease, production of nutritionally improved products, etc.

> In 1997, the first transgenic cow, Rosie, produced human protein-enriched milk (2.4 grams per litre). The milk contained the human alpha-lactalbumin and was nutritionally a more balanced product for human babies than natural cow milk. (p. 213)

Just one argument that goes against the creation and production of transgenic organisms stated in the text can be taken as,

> Genetic modification of organisms can have unpredictable results when such organisms are introduced into the ecosystem. (p. 213)

Since the results are unpredictable, this means that the technology does not guarantee any security and is thus much less than harmless. The prior editions of the same textbook offer a much clearer picture of the misuse of Genetically Modified Organisms.

> Alarmingly, biological techniques are being increasingly misused to produce improved variety of infective agents for using them as bioweapons. Such misuse includes the development of antibiotic-resistant microorganisms with increased infectivity. For example, anthrax is an acute infectious disease caused by the spore-forming bacterium Bacillus anthracis. Spores of B. anthracis can be produced and stored in a dry form keeping them viable for decades in storage or after release. A cloud of anthrax spores, if released at a strategic location to be inhaled by the individuals under attack may act as an agent of effective weapon of bioterrorism. An attack with bioweapons using antibiotic-resistant strains would, thus, initiate the incidence and spread of communicable diseases, such as anthrax and plague, on either an endemic or epidemic scale. (*Biology Textbook for Class XII*, 2002)

From the above quote, it is clear that the technology as sophisticated as genetic modification can be misused for human destruction, and hence raises a dilemma that whether such researches should be pursued or not? Who decides the course of a particular scientific research? And who is responsible for curbing its misuse? The textbook, be it any version, refuses to take any stand on this situation and just discusses the pros and cons.

Ethical Basis: The making and breeding of transgenic animals has recently generated a furore amongst the animal rights activists and people for the welfare of animals. The ethical arguments that are being posed by them include the following:

- Is it ethical to alter and tinker with the genome of any animal for human benefit?
- Does the utilitarian perspective (or cost–benefit analysis) a sufficient assumption for transgenesis of animals?
- Do not we require conducting a need analysis for creating transgenic animals?
- By creating transgenic animals, are we tampering with nature's rules or infringement of certain religious firmaments?
- Ethical treatment of transgenic animals in the labs and inflicting least amount of physical discomfort to them.
- What about the success rate of transgenic animals? Tackling the issue of animal deaths and wastage during such experiments.

All the above-mentioned concerns have been taken into account by certain organizations for animal welfare such as PETA (People for the Ethical Treatment of Animals) and also by ANZCCART (Australian and New Zealand Council for the Care of Animals in Research and Teaching) that abide by the utilitarian perspective in the creation of Transgenic animals (Gott, 2002). Depending upon the cost–benefit analysis of creating transgenic animals, the following situations emerge from case analysis:

- Some uses are generally acceptable, because their costs (usually welfare, health and environment) are negligible.
- Some uses are acceptable because the benefits are sufficient than the minor costs.
- Some uses are intrinsically unacceptable, irrespective of the magnitude of benefits or costs.

Many reports have since then worked upon these principles of utilitarianism with one of them being the Banner Report of 1995, delivered by U.K.'s Ministry of Agriculture, Fisheries and Food. This report explicitly stated that the assessment and resolution of ethical matters simply on the basis of cost–benefit analysis is not acceptable without the acknowledgement of intrinsic objections of some actions. The report summarized the intrinsic objections as,

> "…involves an essentially improper attitude towards them, expressing, in effect, the view that animals are no more than the raw materials for our scientific projects or agricultural endeavours …" Such an attitude "fails to take account of the fact that the natural world in general, and animals in particular, are worthy of our respect as possessing an integrity or good of their own, which we ought not simply to disregard." (p. 12)

However, the invaluable information provided by way of transgenic animals cannot be denied. The following two examples illustrate the benefits derived from such animal models:

Alzheimer's Mouse: Alzheimer's mouse is a creation of scientists designed for the treatment of deadly Alzheimer's disease. The mouse does not suffer much and has given some valuable information about the disease, its occurrence and even shown a hope for its treatment by way of developing vaccines (Schenk et al., 1999) which are now presently in the second phase of clinical trials by Élan Pharmaceuticals.

Oncomouse: It is used to study cancer and development of tumours. Cancer is today, one of those diseases whose treatment is still being studied and researched and the extent of deaths and devastation caused is also enormous. So, a transgenic organism that provides us with the significant information is of great medical utility and for treating millions of humans suffering from the disease. However, the mouse in this case develops multiple tumours with uncontrolled spread leading to pain and suffering to the animal. This has led to the many ethical concerns on the creation and development of such mice.

Thus, each case of transgenic animals needs to be analysed separately for its benefits and probability of suffering and pain to the concerned animal. Also, a need assessment should be done before creating any transgenic organism.

3.4.7 Ethical Issue: Biodiversity and Conservation

Type of ethical issue	Environmental
Category no.	4.2
Broad area	Environmental ecology
Length of issue covered in textbook	1 page
Reference to any ethical theory	Utilitarian
Arguments raised	Positive arguments in the favour of conserving biodiversity
Kind of understanding fostered	Ethical decision-making

The issue being raised here is Biodiversity and the need for its conservation. We all know that Biodiversity is important not just for the sake of aesthetic value that it provides but also for our very existence. The textbook provides valid reasons for it and relies upon the utilitarian ethics to substantiate this viewpoint. The narrowly Utilitarian argument posed in the textbook is,

> Humans derive countless direct economic benefits from nature food (cereals, pulses, fruits), firewood, fiber, construction material, industrial products (tannins, lubricants, dyes, resins, perfumes) and products of medicinal importance. (p. 265)

A broadly Utilitarian argument that has been given in the textbook says,

> …biodiversity plays a major role in many ecosystem services that nature provides. The fast dwindling Amazon forest is estimated to produce, through photosynthesis, 20 per cent of the total oxygen in the earth's atmosphere. Can we put an economic value on this service by nature? (pp. 265–266)

Whereas the following argument in the text presents an ethical stand on the issue:

> The ethical argument for conserving biodiversity relates to what we owe to millions of plant, animal and microbe species with whom we share this planet. Philosophically or spiritually, we need to realize that every species has an intrinsic value, even if it may not be of current or any other economic value to us. We have a moral duty to care for their well-being and pass on our biological legacy in good order to future generations. (p. 266)

The present text tries to convince that there is an inherent need to protect and preserve biodiversity which is regarded as our moral prerogative (Deontological ethics). However, the notions of 'intrinsic value' and 'moral duty' are not being discussed explicitly which could have strengthened the argument further.

Ethical Basis: There is a separate branch of ethics dealing with environment, ecology and the conservation of Biodiversity referred to as 'Environmental Ethics' that refers to the moral status of human being and its relationship to the environment as well as other non-human beings and their interrelationships (Brennan, 2008). There is usually a dilemma as to which and whose ethics should we follow, i.e. the ethics of the environment (Environmental Ethics), the ethics of human behaviour and human benefit (Human Ethics) or the ethics of non-human beings' existence (Non-human/Animal Ethics)? The present textbook, however, does not resolve this conflict; rather it does not even pose it. Two broad notions have emerged out of this environmental ethics, namely, Shallow Ecology and Deep Ecology (Naess, 1973), which are somewhat similar to what is being quoted in the textbook as the Narrowly Utilitarian and the Broadly Utilitarian perspective. The major differences between the two can be given as follows (refer to Table 3.2).

Thus, the outlook portrayed by contenders for Shallow Ecology is purposely 'Anthropocentric', which is narrow and materialistic, whereas that portrayed by advocates of Deep Ecology pertain to the 'Biocentric' view. The Biocentric outlook is more holistic and encompasses the welfare of all living organisms. As per this

Table 3.2 Principles of shallow ecology and deep ecology (Kanungo, 2007)

Shallow ecology	Deep ecology
Natural diversity is valuable as resources for us	Natural diversity has its own intrinsic value
It is nonsense to talk about value except as value for mankind	Equating value with value for humans reveals a racial prejudice
Resource means resource for humans	Resource means resource for living beings
Nature is cruel and necessarily so	Man is cruel but not necessarily so

view, all the organisms have equal rights and no one is superior or inferior, thus promoting equality and justice to all. In order to subsume the inanimate and the nonliving world, a 'Cosmocentric' outlook should be right there in place, hence progressing towards the ethical interrelations with the environment, its resources and biodiversity.

As per the narrowly utilitarian view and the shallow ecological theory, it is mostly the interests of humans that are being held supreme as against those of animal community, which are being regarded as means to serve the mankind in whichever way they can. Be it animal experimentation, medical drug testing or any other such practice for the benefit of the larger human community, the issuc being raised here by many animal rights activists is the fairness of conducting experiments and medical trials on animals. This can still be justified on the basis of a utilitarian ethics of saving the lives of innumerable humans and enhancing the quality of living. But indiscriminate use of animals for shoes, belts, handbags, saddles, luggage or food is not justified where we know that an alternative exists.

In order to resolve the conflict between rights and values of humans and non-human beings, Peter Singer[7] observes,

> ...That humans and animals are utterly different kinds of beings was unquestioned for most of the course of western civilization. The basis of this assumption was undermined by Darwin's discovery of our origins and the associated decline in the credibility of the story of our divine creation in the image of God. Darwin himself argued that the difference between us and animals is one of degree, rather than of kind- a view that even today some find difficult to accept. (p. 64)

Thus arguing for equal rights and value of life for humans and non-human beings by applying the principle of sentience and feeling of pain and distress which is similar across the entire animal kingdom (Singer, 2011).

3.4.8 Ethical Issue: Environmental Pollution

Type of ethical issue	Environmental
Category no.	4.3
Broad area	Environmental ecology
Length of issue covered in textbook	15 pages
Reference to any ethical theory	None
Arguments raised	Deleterious impact of pollutants on the atmosphere as well as case studies for remedying the harmful effects of those pollutants
Kind of understanding fost`ered	Conceptual- and application-based

[7]"Equality for Animals", in Practical Ethics by Peter Singer, p. 64.

The present account of Environmental issues being raised in the textbook has an explicit focus on explaining facts about various kinds of environmental pollutants, such as air pollution, water pollution, solid wastes, radioactive wastes, global warming, Ozone depletion, deforestation, etc. However, along with the detailing on the different aspects of each and every kind of pollution, the textbook stresses upon ways of curbing it by suggesting some alternative measures in a separate section namely, "Case Studies" (pp. 277, 279, 280, 284). Here, the one case study of "People's participation in conservation of forests (p. 284)" narrates a story of courage and emotion exhibited by the people of Bishnoi community who loved and regarded their trees even more than their own lives.

Ethical Stance: The entire debate is centred on curbing ever-surging levels of pollution and climate change on one side and development on the other, which appears to be a difficult choice especially for developing countries. The developed nations whose carbon footprint comes out to be maximum still the onus of sharing the burden lies equally with both the developed and the developing countries, which is unfair. Another question that arises is whether a development at the cost of environmental loss and climate change are desirable? Can there be a balance between the two, so as to find out ways by which technology does not harm the environment, or if it does to a minimal extent, then the benefits derived should be several times more than the cost incurred? Another point which can be put here is that despite governmental concerns and policymaking, what are some of the chief responsibilities of the citizens of this country towards reducing this climate change? Can individuals bring about a perceptible reduction in carbon emissions and contribute towards global environmental sustainability? These are the questions that the present textbook needs to address.

The textbook suggests some alternative procedures such as the use of CNG to curb air pollution, sewage treatment plants for the purification of polluted water, e-waste treatment, etc., but can these be taken as solutions to the problem or are merely handling the outcomes of it? Can CNG replace the requirement for diesel and oil everywhere? Can sewage treatment plants treat all the impurities present in water? The point that emerges from all the above discussions is the creation of a balance, such that the level of pollution or adulteration can be minimized at the level of its production, which requires efficient planning and implementation of a set of rules and norms. Hence, taking from a utilitarian perspective, the cost–benefit analysis should be done with the active participation of people from all walks of life, so as to arrive at a consensus decision, thus considering the ethics of justice which has a predominance for subsistence rather than luxury (Singer, 2011 p. 228).

Again, the ethical point of view is towards reduction in materialistic pursuits, leading a life of simplicity. The following statements conjure us to think that way,

> An environmental ethic rejects the ideals of a materialist society in which success is gauged by the number of consumer goods one can accumulate. Instead it judges success in terms of the development of one's abilities and the achievement of real fulfilment and satisfaction. It promotes frugality and re-use, insofar as that is necessary for minimizing the impact we have on the planet. (Singer, 2011, p. 254)

3.5 Conclusion

The critical analysis of the K-12 Biology textbook was undertaken with the chief aim of highlighting the place of ethics in Science curriculum in Indian schools. The major finding that emerges from the analysis is that the ethical issues are only scantly listed and discussed within the text at some points where the central focus is on technology and its benefits. The conflicting arguments and controversies that surmount these technologies have largely been ignored in the present text in order to maintain objectivity. There are many ethical issues associated with animal experimentation, cloning and human genomic project that the present textbook of Biology completely neglects. However, such a notion is against the democratic ideal of a curriculum and favouring the dominant form of knowledge. According to Apple (1993), textbooks disseminate only particular constructions of reality that are deemed to be 'legitimate' for a particular society by selecting and organizing from a vast repertoire of knowledge. Such a text builds only one-sided view of technology rather than a plethora of interpretations of issues and concerns attached to these technologies. This limited view of technology presented in the textbook not only obfuscates the holistic vision but also decries education of its true purpose as stated by Apple and Beane (2006),

> that young people should not take on social issues because they are not ready to understand the complexities involved or because they might become depressed. However, such arguments ignore the fact that young are real people living out real lives in our society; many of them know all too well, from their own lived experiences, about the consequences of racism, poverty, gender bias, homelessness and so on. Such arguments simply seek to avoid the possibility that young people might come to see the political, ethical, and social contradictions that detract from their own dignity and seek to act against them.

Thus, in order to build such an educational experience that favours multiplicity of opinions along with a critical reflection on the different socio-ethical issues inherent in a particular technology and its application, a 'thematic curriculum' is required that engages the learners with their real-life problems and issues (Beane, 1993 cited in Apple & Beane, 2006). The present curriculum of science aims only to create a functional literacy and participation in research and development-related endeavours but does not foster a critical literacy that enables the growth of a genuine understanding and active participation in all social spheres of life. In order to foster such a holistic understanding, ethical issues deserve a much coveted place within the Science curriclum.

References

Apple, M. W. (1991). The culture and commerce of the textbook. In M. W. Apple & L. Christian-Smith (Eds.), *The politics of the textbook* (pp. 22–40). New York: Routledge.

Apple, M. W. (1993). *Official knowledge: Democratic education in a conservative age*. New York: Routledge.

Apple, M. W., & Beane, J. A. (2006). *Democratic schools: Lessons from the chalk face*. New Delhi: Eklavya Publication.
Bernstein, B. B. (1971). *Class, codes and control, Vol. 1: Theoretical studies towards a sociology of language*. Routledge & Kegan Paul Publics.
Brennan, A. (2008). Environmental ethics. In *Stanford encyclopedia of philosophy*. Retrieved from http://plato.stanford.edu/entries/ethics-environmental/.
Cohen, L., Manion, L., & Morrison, K. (2007). *Research methods in education*. New York: Routledge Publications.
Dhar, B., & Saha, S. (2014). An assessment of India's innovation policies. Discussion Paper 189. Research and Information System for Developing Countries, New Delhi, India.
Gott, M. (2002). Ethical issues relating to transgenic animal production. *The Australian and New Zealand Council for the Care of Animals in Research and Teaching*, *15*(1), 1–14. Retrieved from http://www.adelaide.edu.au/ANZCCART/news/15_1.pdf.
Hinman, L. H. (2009). *Abortion: An overview of the ethical issues*. Retrieved from http://www.nfaw.org/social_policy/papers.html.
ICMR (2004). Regulatory regime for genetically modified foods. The Way Ahead (A Study). New Delhi, India.
Kanungo, P. D. (2007). *Perspectives in applied ethics*. New Delhi: Books India International.
Knight, A. (2009). Perceptions, knowledge and ethical concerns with GM foods and the GM process. *Journal of Public Understanding of Science, 18,* 177–188.
Krippendorp, K. (2004). *Content analysis: An introduction to its methodology*. Thousand Oaks, CA: Sage.
Levinson, R., & Reiss, M. J. (2003). Ethics and education: Issues and scenarios. In R. Levinson & M. Reiss (Eds.), *Key issues in bioethics: A guide for teachers* (pp. 3–13). London: RoutledgeFalmer.
Lotz, M. (2005). *Ethical arguments concerning the moral permissibility of abortion*. Retrieved from http://www.nfaw.org/social_policy/papers.html.
Lynch, P., & Paul, D. S. (1985). Ten decades of the science textbook: A revealing mirror of science past and present. *Journal of Science and Mathematics Education in Southeast Asia, 8* (2), 31–42.
Naess, A. (1973). The shallow and the deep, long-range ecology movement. *Inquiry, 16,* 151–155.
Pascalev, A. (2003). You are what you eat: Genetically modified foods, integrity and society. *Journal of Agricultural and Environmental Ethics, 16,* 583–594.
Position Paper National Focus Group on Curriculum, syllabus and textbooks. (2006). New Delhi: NCERT.
Schenk, D., Barbour, R., Dunn, W., Gordon, G., Grajeda, H., & Guido, T. (1999). Immunization with Amyloid-Beta attenuates Alzheimer-disease-like pathology in the PDAPP Mouse. *Nature, 400,* 173–177.
Singer, P. (2011). *Practical ethics* (3rd ed.). Cambridge: Cambridge University Press.
Tadiar, F. M., & Robinson, E. T. (1996). Legal, ethical and regulatory aspects of introducing emergency contraception in Philippines. *International Family Planning Perspectives, 22*(2), 76–80.
Weber, R. P. (1990). *Basic content analysis* (2nd ed.). Thousand Oaks, CA: Sage.
WHO (2000). Safety aspects of genetically modified foods of plant origin. Report of a Joint FAO/WHO Expert Consultation on Foods Derived from Biotechnology. WHO Headquarters, Geneva, Switzerland.
Woolfolk, A. (2004). *Educational psychology* (9th ed.). Delhi, India: Pearson Education Inc.

Chapter 4
Pedagogical Approaches to Ethical Issues in Biosciences

Abstract The present chapter draws interrelationship between the concept of pedagogy (along with pedagogical content knowledge) and the subject matter of ethical issues in Biosciences for developing an ethical awareness and understanding among the learners for these issues. Pedagogy acts as an axle to develop this reasoning about the ethical issues in Biosciences that is reflected in different approaches that teachers adopt in addressing these issues in their classrooms. In order to develop this pedagogical reasoning the construct, Pedagogical Content Knowledge (PCK) serves as a criterion. This category of teachers' knowledge in the area of ethical issues in Biosciences needs to be enhanced and enriched for an effective classroom transaction and for generating an authentic ethical understanding among the learners. The present chapter elaborates upon the concept of pedagogical content knowledge and its ramifications into the different teaching learning approaches for transacting ethical issues in Biosciences.

Keywords Pedagogy · Pedagogical content knowledge (PCK) · Ethical understanding · Biosciences · Ethical issues

4.1 Concept of Pedagogy

Pedagogy is referred to both as a science and an art, as enunciated by a number of educational philosophers such as Kant (1724–1804) and Hegel (1770–1831). Both Kant and Hegel added an altogether higher dimension to the understanding of 'pedagogy' per se. It was earlier understood to be within discipline-specific boundaries but was not viewed from the perspective of moral development. Kant stressed a lot on institutions of education and their organizers as harbingers of moral and character development of individuals towards a meaningful living and responsible citizenry. Pedagogy should be such that it promotes the development of the natural aptitudes inherent in human nature. Kant relates pedagogy to its emancipatory role by way of critical philosophy which ascribes to the cultivation of reason, which should be mastered by those who are to teach. This particular notion of Kant coincides with the critical pedagogy of the neo-Marxists and post-Marxists.

A. Saxena, *Ethics in Science*, https://doi.org/10.1007/978-981-32-9009-9_4

Pedagogy can be taken both as a cognitive and a conative aspect of a teacher's personality and has more to do with the 'pedagogical reasoning' (Wilson, Shulman, & Richert, 1987, p. 118). Pedagogical reasoning is a way of thinking that transforms the subject knowledge of the teacher into forms that can be readily understood by the students. Thus, the idea of 'pedagogy' resides in a teacher's mental processes and is executed in the manner in which she transacts a particular lesson in the classroom. The different categories of teachers' professional knowledge repertoire as proposed by Shulman (1986) can be taken as follows:

i. Subject matter content knowledge,
ii. General pedagogical knowledge,
iii. Pedagogical content knowledge,
iv. Knowledge of Educational Aims,
v. Curricular Knowledge,
vi. Knowledge of the learners,
vii. Knowledge of other content.

However, the major focus of research has been on the first three components of Shulman's classification (Baumert & Kunter, 2006). Besides subject matter content knowledge, it is the general pedagogical knowledge and pedagogical content knowledge that is at the heart of successful teaching. *Ongoing research in this area defines pedagogy as a component of teacher's knowledge that needs to be acquired by the novices in the field. A novel term for such a knowledge construct was coined by L. S. Shulman as,* ***Pedagogical Content Knowledge (PCK)*** *in 1986 in one of his seminal papers, 'knowledge and teaching'.* This sprouted a new area of research for teachers, teacher educators, researchers and other academia.

4.2 Pedagogical Content Knowledge (PCK)—An Evolving Construct

The different facets of a teacher's knowledge base subsumed under one construct of Pedagogical Content Knowledge (PCK) was for the first time researched and reported by L. S. Shulman in the year 1986 in his famous work—"Those Who Understand: Knowledge growth in teaching" as well as in another famous work in the year 1987—"Knowledge and Teaching: foundations of the new reform". Defining PCK in his own words:

> The category of PCK includes – the most regularly taught topics in one's subject area, the most useful forms of representation of those ideas, the most powerful analogies, illustrations, demonstrations, examples, explanations, and ways of representing and formulating the subject that makes it comprehensible to others. PCK also includes an understanding of what makes the learning of specific topics easy or difficult; the conceptions and preconceptions that students of different ages and backgrounds bring with them to the learning of those most frequently taught topics and lessons.

However, since the introduction of the term PCK and the meaning connoted to it, there have been very many debates among those in the educational fraternity including teachers, researchers, educational psychologists and curriculum developers. They regard the notion of PCK as highly structured and deterministic, leading to its many modified versions. Thus, it becomes imperative to understand the debates centred on this construct PCK and generate a viable and concurrent understanding by studying and analysing various related theoretical models that have emerged from it.

The reason for referring to the construct PCK in the present book is to use it as a tool to understand the different facets of teachers' knowledge base with respect to ethical issues in Biological Sciences.

Research has proven that teacher's knowledge has a significant impact on enhancing the quality of learning as well as cognitive activation (Clermont, Krajcik, & Borko, 1993; Van Driel, Verloop, & De Vos, 1998; Ergonenc, Neuman, & Fischer, 2014).

4.3 Ethical Issues in Science Curriculum

The importance of addressing ethical issues in Biological Sciences cannot be overemphasized and many researchers have delineated their role in enhancing scientific learning and scientific literacy among students (Cross & Price, 1996; Pedretti, 2003; Zeidler, Sadler, Simmons, & Howes, 2005). The curriculum of science needs to be made more open and interdisciplinary so as to incorporate these issues with their true intent and spirit. At present after having reviewed the curriculum of Biological Science at the senior secondary level, these issues only make a fleeting presence within the textbook and the syllabus also does not have any mention about them. The teachers hesitate rather abstain from discussing these issues in the classroom due to their contentious nature and since they require digressing from the conventional methods of teaching, which at times puts a teacher's own position in jeopardy (Hodson, 2003; Hughes, 2000). The objective of writing the present chapter is to review recent literature on strategies for dealing with ethical issues in the science classroom.

4.4 Dealing with 'Ethics'—The Deweyan Way

A Deweyan approach to science teaching is said to be favouring 'ethics' education as reported in some studies (Webster, 2008) according to which there is no need for a science teacher to acquire any special expertise for dealing with these ethical issues in the classroom. For this to happen, it needs to be recognized that science and ethics should no more be regarded as disjoint spheres of knowledge rather both guide each other and are necessary educative components for a scientific and moral development of an individual. The point of intersection between science and values according to Dewey occurs when it comes to the application of science and the consequences therein (Dewey & Childs, 1989, p. 78). Science education or for that matter any education should liberate an individual from the state of passivity towards active role bearing, learning to take decisions for themselves rather relying upon the authority and governance, and should provide a holistic learning experience rather than mere transmission of knowledge (Dewey, 1975). In fact, Dewey argues that scientific mode of inquiry should also be applied to value judgments, which our science teachers can easily facilitate and initiate among the students while dealing with questions of ethics, morals and religion. These areas should not be regarded as outside the realm of scientific inquiry or critical thought (Dewey, 1988, p. 71). The Deweyan philosophy of science education does not specifically entail the teaching of specific issues rather towards a democratic process, wherein the views of each and every student are acknowledged and taken into consideration.

4.5 Ethical Issues in Science: Pedagogical Considerations

The contentious nature of these issues raises concerns over their pedagogy and the teaching–learning processes so as to generate an authentic understanding among students. Some of the models such as one proposed by Burnham and Mitchell (1992) include five stages in order to reach an ethical conclusion, viz., observation, questioning and hypothesizing, information gathering, analysis and ethical deliberation, and ethical decision-making. Knowledge about students' own worldviews also plays an important role in determining the kind of ethical conclusion arrived at and a teacher should try to inquire into it and share her own opinion as well with the students (Oulton, Dillon, & Grace, 2004). As Lynch and Mcknena (1990) also proposed that the worldviews of different individuals vary and are built up right from the time of birth and develop in the course of responding to innumerable social and cultural interactions with the world. Oulton et al. (2004) has propounded some salient points that need to be kept in mind while dealing with controversial issues, such as, multiple viewpoints surrounding these issues need to be regarded well; different ways of interpretation occur depending on different worldviews, values, etc.; and an issue can only be resolved once more information becomes available.

4.5.1 Contextualizing the Teaching of Moral and Ethical Issues

Case-based approaches have also been advocated for basing the scientific concepts into real-life events and happenings. A discussion can be generated in the classroom among the students about the possible impact of a particular research or divulsion of an inherent fallacy or myth in the arguments generated for finding out the reasons and logistics behind it. Sometimes the issue is debatable on moral or ethical grounds as and when all the arguments appear to be equally convincing on grounds of human health, safety, well-being and growth and development of human society. In such a case, the person can draw on from Giligan's and Kohlberg's theories of moral development and reasoning that help in determining the moral or ethical stance that a person is likely to take depending upon the circumstance (Keefer, 2003). These ethical modes of enquiry and the theories that guide towards an ethical decision-making dovetail with the problem-based learning approach and inquiry approach in Sciences as there happen to be instances where one problem can have a number of solutions each affected by factors lying outside the subject's own boundaries, such as sociopolitical context, religion, gender, etc. Hence, the role of moral reasoning in selecting the best alternative can be justified here.

4.5.2 Contrived Learning Experiences for Teaching Ethical Issues in Biosciences

Many theorists and pedagogues have suggested different means of overcoming the difficulty of teaching ethical issues by the creation of artificial contexts in the form of role plays, or viewing a movie/play/documentary by making use of multimedia. Role plays are active forms of learning where the students themselves become a part of a contrived situation focussed on a particular issue, here they are themselves a witness to the ethical dilemmas, are able to explore other people's perspectives and apply their own ethical judgment and provide solutions by analysis of all the arguments (Oulton et al., 2004, p. 76). In case of a documentary or a movie, the role of students is to some extent restricted as then they are passive viewers, however, the situation can be modified by asking students to look for particular themes and points in the movie that is of ethical significance and later, a discussion in the class can be generated based on those marked points or instances. In both cases, the role of the teacher is questionable, as what role are teachers subjected to in such a case? Should we completely negate the role of lecturing as a teaching method? Here, a midway route has been suggested such that in dealing with the ethical theories and different modes of ethical enquiry and problem-solving lecture method proves to be useful and effective, however, long periods of lecture can be interspersed with short intervals of reflection and silent thinking or just pondering over the issue (Wilmott & Bryantt, 2005).

4.5.3 Situated Learning Approach for Transacting Ethical Issues in Biosciences

Due to the growing concern for the place of these ethical issues within the science curriculum, there has been an increasing emphasis on teaching scientific content by relating it to the daily-life issues encountered by the learners. This has given rise to a 'situated learning' approach to Biology teaching (Gilbert, 2006; Van Aalsvoort, 2004). Such an approach is also said to improve student motivation (Sadler, 2009). Researchers have shown that science teachers find themselves as unfamiliar and uncomfortable dealing with socio-scientific issues and putting them across to the students by way of classroom discussions (Day & Bryce, 2011; Levinson & Turner, 2001; Osborne, Erduran, & Simon, 2004). In such cases, teachers may adopt different stances while dealing with ethical issues in the classroom as suggested by some researchers. They may acquire a neutral position, also referred to as 'procedural neutrality' while conducting debates (Stenhouse, 1983 cited in Oulton, Dillon, & Grace, 2006) or 'committed impartiality' proposed by Kelly (1986) where the teacher provides all sides of an argument but also shares her own views with the class.

Wieringa, Janssen and Van Driel (2011) conducted a research to establish the possibility of '*Science theatre at school*' for providing a context to learn about socio-scientific issues, especially the ethical and moral implications embedded within these issues. The method involved staging the play with life characters portraying real-life contexts and situations so that the audience can relate it with their own day-to-day experiences. In this particular study, the subject chosen was 'food science and technology' and various roles had been assigned to different actors such as snack man fast-food director, scientists, the naked genome, which is the role of a youngster taking care of himself. The sample included some of the students (20 boys and 19 girls) and 11 experts in the fields such as science communication and education, theatre studies, psychology, philosophy and nutrigenomics. The data collection involved taking in student views from the focussed group discussions after the play and semi-structured interviews with the experts. There appeared a stark variation in the nature and inclination from the student responses as some were positively influenced by the idea of theatre in science, which motivated them to study and understand science, while for others, it was a compulsory task that had to be attended and did not serve much of the purpose in terms of understanding the issue and according to them, many concepts just remained at the superficial level of understanding.

In yet another study (Ritchie, Tomas, & Tones, 2011), a novel approach has been tried and experimented upon in order to deal with the socio-scientific–ethical issues in the secondary science classrooms by way of developing '*Biostories*'. The study adopted a quasi-experimental design wherein two groups were made—one *treatment* group and another a *comparison* group. However, an experienced Biology teacher who taught Biology to both the groups was common, with the only change in the pedagogy adopted for the teaching of the treatment group. In the treatment group, the students were asked to develop Biostories on a given topic of

Biosecurity by converting the technical subject knowledge given in their textbooks into conversational prose to be understood by a layman. The results of the study were positive in the sense that students in the treatment group showed greater interest in science, along with better familiarity with and understanding of Biosecurity-related concepts as compared to the comparative group. There was also an improved derived sense of scientific literacy in the biostories of the students in the treatment group. The study advocates the introduction and provision for 'narratives (biostories)' on suitable SSI within the science curriculum at all levels to foster the development of scientific literacy among citizens.

4.5.4 Role of Teachers' Pedagogical Content Knowledge in Transforming Learner's Experiences

In a study conducted by Van der Zande, Akkerman, Brekelmans, Waarlo and Vermunt (2012) on 'Expertise for Teaching Biology situated in the context of Genetic Testing', a sample of nine experienced school teachers was taken and were examined for their expertise of teaching genetics in a situated context of 'genetic testing'. The various dimensions of teachers' expertise that were being studied included knowledge in pedagogical content, the interpersonal and moral expertise areas concerned with the teaching of genetics. The teachers were interviewed on a semi-structured interview scale and the classes of five of them were video recorded and stimulated recall interviews were taken. Data was analysed by categorization and coding of responses using different schemes. The results indicated that all the case teachers used a variety of methods (mixed-method approach) in order to deal with these controversial socio-scientific issues, such as, stories, multimedia, discussions ranging from teacher-centred (chalk and talk) to entirely student-centred (think-, pair-, share- and problem-based small groups), whole-class activities as well as group work and individual assignments, a variety of practical lessons and assignments for presentation (poster), evaluation (writing assignments) or reflection (yes/no interlude). Most of them also had good interpersonal skills, however, within the area of moral expertise, most of the teachers (seven out of nine) underscored and exhibited difficulties in dealing with the moral dilemmas within these issues and their inability to address them appropriately to the students. They admitted about having no specific teaching and learning activities to deal with this aspect of learning at their disposal. However, they all agreed that they require assistance in this regard to strengthen students' argumentation vis-à-vis these socio-scientific issues.

4.6 Socio-Ethical Issues and Changes in Students' Understanding

The objective behind teaching ethical issues in the classroom is not merely awareness about them, rather transforming students' understanding and a deeper engagement with the controversies and debates therein. The development of skills of critical inquiry is also important in this regard, as Crick writes, 'students ask more awkward questions, hopefully in a more sensible way, and not be put off by stock answers' (Crick, 2001, p. 34). A science classroom with the preponderance of ethical issues is said to transform students' conceptual level understanding beyond the mere recital of scientific facts and theories by enlivening the experience of studying science. An open discussion on the varied aspects of these issues helps in the epistemological growth of the discipline as a whole and removes sciences from the elitist tag that has always been a barrier for its contextualization. This will also help in unearthing certain traditional and indigenous ways of thinking and practicing resulting in the promulgation of a multicultural view of science education. Such a paradigm shift in the curricular approach of teaching science connects to the students' value systems, sense of ethics and moral reasoning (Zeidler & Sadler, 2008).

4.6.1 Enhancing Scientific Argumentation Skills Through Online Discussions

In a recent study carried out by Wishart, Green, Joubert and Triggs (2011), a novel pedagogy for ethical issues based on online and face-to-face discussions has been suggested. The study was conducted in the U.K. and aimed at initiating teachers towards discussions on socio-scientific and ethico-moral issues related to sciences particularly Biology, since they are perceived to be one of the difficult areas for teaching. The researchers in this study planned three different kinds of discussions —(i) Online Within the School Discussion; (ii) Online Between the Schools Discussion; and (iii) Face-to-Face Discussion. The quality of responses and arguments gathered from these discussions were analysed based on a five-level grid developed by Erduran, Simon and Osborne (2004). The sample for the study included six teachers as co-researchers from the schools across England, along with the student sample of 84 students who were all year 12 students (aged 16–17) of the above-mentioned teachers. The schools had the computer and Internet facility, which was an important requirement for this study. The results of the study indicated that students felt more comfortable with online form of discussion as compared to face-to-face, and tend to post more rebuttal arguments as compared to face-to-face where the arguments tend to reach the consensus soon. But, each mode of discussion has its own advantages, such as, with face-to-face discussions, the students could make out from the facial expressions and other gestures whether other students are really agreeing on a point or not. The implications that can be drawn from such a study is that besides regular teaching–learning in the classroom some time should be spared for initiating such discussions so as to enable the students in developing an argumentative mindset related to ethical issues in Biological Sciences.

4.6.2 Development of Ethical Subjectivities Among Learners

A study conducted by Bazzul (2015) emphasizes that the presentation of ethical issues in the textbook influences the kind of '*Ethical Subjectivities*' developed among the teachers and students. This particular study was carried out on four Ontario textbooks of high school Biology and looked at the kind of ethical questions and exercises given in those textbooks. The study devised some ways of addressing and presenting ethical issues within the textbooks so as not to favour any one particular ethical position or decision-making rather providing open-ended choices to the larger public.

4.7 Developing an Alternative Framework for Addressing Ethical Issues in Biosciences

Drawing from the theoretical orientation, it appears that a 'critical pedagogy' approach, which addresses the desired goals for science teaching and especially, teaching of ethical issues in Biological Sciences needs to be evolved. In order to understand these issues, a critical mindset is required which can be fostered only if teachers promote dialogue and debate in their classrooms by designing problem-based exercises (Freire, 1972); questioning the existing structures of knowledge (Apple & Beane, 2006); adopting context-specific pedagogy (Giroux, 2011); and addressing the accompanying ethical and moral responsibilities (Kincheloe, 2005). Thus, a critical understanding of ethical issues and a critical pedagogy approach for dealing with these issues in the classroom is required for building a sound Pedagogical Content Knowledge of teachers.

A relationship between critical thinking about ethical issues, critical pedagogy and Pedagogical Content Knowledge (PCK) can be sought. This can also be interpreted as PCK (of ethical issues) being enriched and improved by a critical understanding of the ethical issues as well as a critical pedagogy approach. Here PCK is constituted by interplay of different knowledge components that are variable and ever dynamic. Enlisting some discrete components of teachers' knowledge is not sufficient to bring about a transformed teaching–learning experience, but thinking and reflection are important to make teaching effective and liberating. In the case of ethical issues, an even higher level of thinking is required

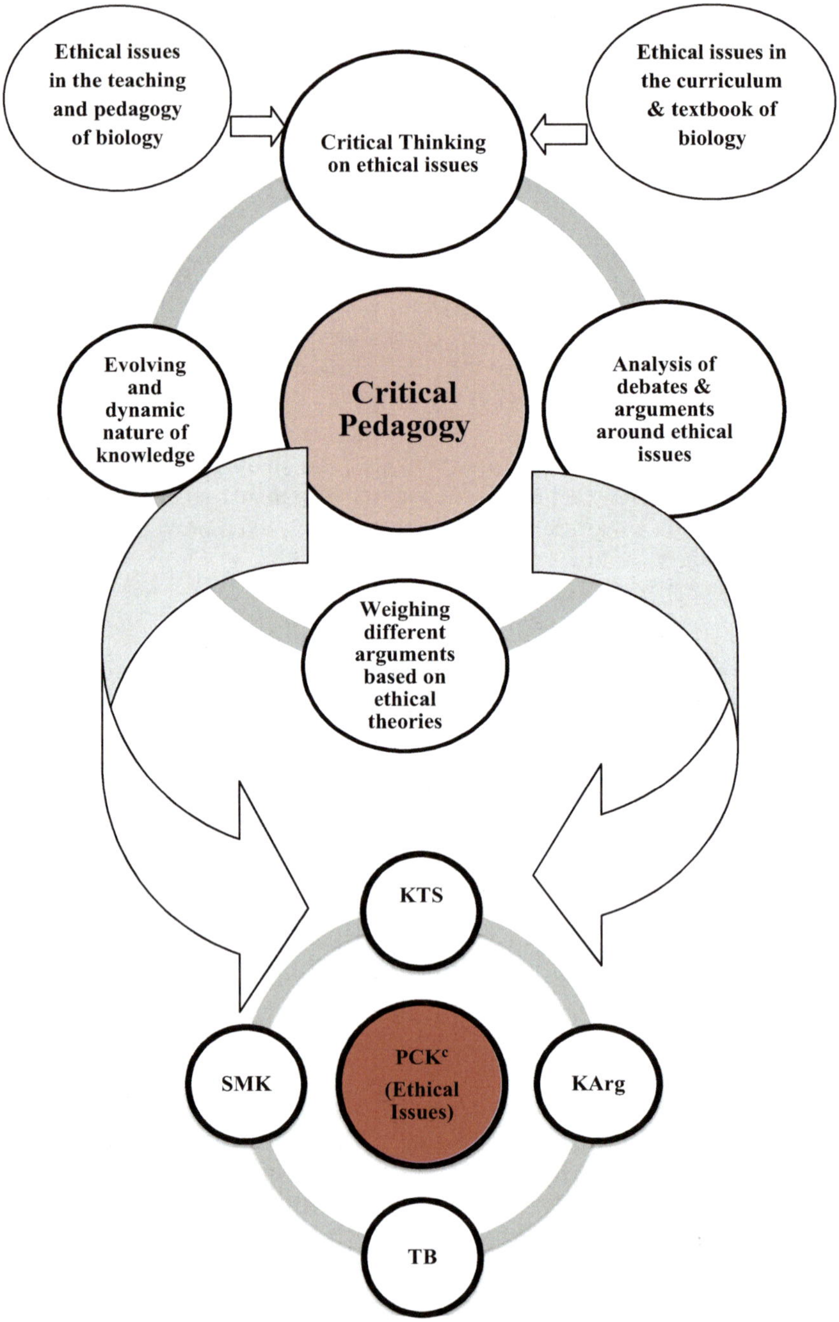

Fig. 4.1 Representation of critical pedagogical content knowledge (PCK^C) (*Source* Author)

which is 'critical thinking' as there is no one method, answer, argument or solution to resolve ethical issues, but multiple and at times conflicting. This means that a teacher needs to think critically about these issues both epistemologically and pedagogically before formulating objectives for the lesson. She needs to be critical about the subject content that she chooses to teach, the transactional strategy that she adopts and the kind of assessment in the form of exercises and tasks given to the students.

A conceptualization of critical Pedagogical Content Knowledge (PCK^{C}) that has emerged based on the present understanding may be depicted as follows (refer to Fig. 4.1).

4.8 Conclusion

It emerges from the above discourse, that there is no single pedagogy to deal with the ethical issues in Biological Sciences. The pedagogical approaches suggested by different researchers in the field are rather unconventional and defy the traditional methods of teaching-learning. The underlying notion is to relate the content with the context of the learner to make it more comprehensible. Since these ethical issues and ethical dilemmas are not being faced by the learners every time in their day-to-day life, therefore, a contrived experience works best here, such as by making use of Biostories (Ritchie et al., 2011); context-based lessons and Science theatre (Wieringa et al., 2011); and online and face-to-face discussions (Wishart et al., 2011).

Each of these approaches is learner-centric and takes into account students' opinions and voices and thus prepare for a democratic classroom. These kinds of teaching–learning practices conform to the critical pedagogy of Freire (1972) and Giroux (2011). The construct of PCK is limiting in the sense that it does not take cognizance of the nuances of critical pedagogy and is rather a more generalized concept. Hence, in the present study, we do not keep ourselves limited to the conceptual understanding of the concept of PCK rather incorporate certain elements of critical pedagogy within its ambit. Keeping in mind the ultimate goal of education, which is towards social reconstruction (Dewey, 1975) and emancipation of society, we situate the concept within the present research domain.

The narrative starts building from the very notion of 'pedagogy' and 'critical pedagogy' further taking into account the evolution of the concept of PCK and its varied dimensions. The construct PCK is not an all-encompassing concept, and the different conceptualizations of the construct given by researchers further develop from it making it a dynamic and an ever evolving field. There are some inherent limitations within the concept of PCK such as, although it has evolved from the behaviouristic notion of teaching, it is a highly cognitive construct, giving least importance to practice. Then, it neglects the role of teachers' emotions or the affective component in teaching–learning, and most importantly, it is delinked with the outcome of teaching in terms of students' learning or understanding. Thus, there is a need to rebuild the present understanding of the construct PCK with its ramifications in the area of ethical issues in biological sciences. This has been taken up so as to reveal and understand the ways in which teachers negotiate these ethical issues in the classroom. A detailed analysis of the recent Biotechnology techniques using the ethical matrix suggests that these technologies are indeed beset with many ethical concerns that need to be taken care of. Ethical issues in science are in itself a contested area of study and their addressal in the classroom is reported to be difficult by the teachers.

A review of studies presented herein gives an idea about the kinds of teaching–learning strategies adopted by teachers and researchers in the classroom to deal with these issues with their students such as role plays, face-to-face and online discussions, designing context-based lessons, biostories. However, there is no mention in these studies about teachers' beliefs and teachers' pedagogical content knowledge with respect to ethical issues and their own ways of dealing with them. An attempt is being made to link the teachers' PCK with respect to the teaching of ethical issues, so as to deliberate and inform about their practice to give a real picture of the situation in the classroom. Also, this will help in identifying the lacunas in the prevalent practice and offering solutions by way of educational implications for the pre-service and in-service teacher education programmes.

References

Apple, M. W., & Beane, J. A. (2006). *Democratic schools: Lessons from the chalk face*. New Delhi: Eklavya Publication.

Baumert, J., & Kunter, M. (2006). Stichwort: Professionelle Kompetenz von Lehrkräften [Cue: Professional competence of teachers]. *Zeitschrift für Erziehungswissenschaft, 9*(4), 469–520.

Bazzul, J. (2015). Tracing "Ethical Subjectivities" in science education: How biology textbooks can frame ethico-political choices for students. *Research in Science Education, 45,* 23–40.

Burnham, M., & Mitchell, R. (1992). Bioethics: An introduction.

Clermont, C. P., Krajcik, J. S., & Borko, H. (1993). The influence of an intensive in service workshop on pedagogical content knowledge growth among novice chemical demonstrators. *Journal of Research in Science Teaching, 29,* 471–485.

Crick, B. (2001). Citizenship and science; science and citizenship. *School Science Review, 83,* 33–38.

Cross, R. T., & Price, R. F. (1996). Science teachers' social conscience and the role of controversial issues in the teaching of science. *Journal of Research in Science Teaching, 33*(3), 319–333.

Day, S. P., & Bryce, T. G. K. (2011). Does the discussion of socio-scientific issues require a paradigm shift in science teachers' thinking? *International Journal of Science Education, 33* (12), 1675–1702.

Dewey, J. (1975). *Moral principles in education.* Carbondale and Edwardsville: Southern Illinois University Press.

Dewey, J. (1988). Human nature and conduct. In J. A. Boydston (Ed.), *John Dewey: the middle works, 1899–1924.* Carbondale and Edwardsville: Southern Illinois University Press.

Dewey, J., & Childs, J. L. (1989). The underlying philosophy of education. In J. A. Boydston (Ed.), *John Dewey: The later works 1925–1953* (Vol. 8). Carbondale and Edwardsville: Southern Illinois University.

Ergönenç, J., Neuman, K., & Fischer, H. (2014). The impact of pedagogical content knowledge on cognitive activation and student learning.

Freire, P. (1972). *Pedagogy of the oppressed.* USA: Penguin Books.

Gilbert, J. (2006). *Science education: Major themes in education.* New York: Routledge Publications.

Giroux, H. A. (2011). *On critical pedagogy.* New York: Continuum International Publishing.

Hodson, D. (2003). Time for action: Science education for an alternative future. *International Journal of Science Education, 25*(6), 645–670.

Hughes, G. (2000). Marginalization of socioscientific material in science-technology-society science curricula: Some implications for gender inclusivity and curriculum reform. *Journal of Research in Science Teaching, 37*(5), 426–440.

Keefer, M. (2003). Moral reasoning and case-based approaches to ethical instruction in science. In D. L. Zeidler (Ed.), *The role of moral reasoning on socioscientific issues and discourse in science education.* Science & technology education library (Vol. 19). Dordrecht: Springer.

Kelly, T. (1986). Discussing controversial issues: Four perspectives on the teacher's role. *Theory and Research in Social Education, 14,* 113–138.

Kincheloe, J. L. (2005). *Critical pedagogy primer.* New York: Peter Lang Publishing Inc.

Levinson, R., & Turner, S. (2001). *Valuable lessons: Engaging with the social context of science in schools.* London: The Wellcome Trust.

Lynch, D. J., & Mcknena, M. C. (1990). Teaching controversial material: New issues for teachers. *Social Education, 54*(5), 317–319.

Osborne, J., Erduran, S., & Simon, S. (2004). Enhancing the quality of argumentation in school science. *Journal of Research in Science Teaching, 41*(10), 994–1020.

Oulton, C., Dillon, J., & Grace, M. (2004). Reconceptualising the teaching of controversial issues. *International Journal of Science Education, 26*(4), 411–423.

Oulton, C., Dillon, J., & Grace, M. (2006). Reconceptualising the teaching of controversial issues. In J. Gilbert (Ed.), *Science education: Major themes in education.* London: Routledge.

Pedretti, E. (2003). Teaching science, technology, society and environment (STSE) education: Preservice teachers' philosophical and pedagogical landscapes. In D. Zeidler (Ed.), *The role of moral reasoning and socioscientific discourse in science education* (pp. 219–239). Dortrecht, The Netherlands: Kluwer.

Ritchie, S. M., Tomas, L., & Tones, M. (2011). Writing stories to enhance scientific literacy. *International Journal of Science Education, 33*(5), 685–707.

Sadler, T. D. (2009). Situated learning in science education: Socio-scientific issues as contexts for practice. *Studies in Science Education, 45*(1), 1–42.

Shulman, L. S. (1986). Those who understand: Knowledge growth in teaching. *Educational Researcher, 15*(2), 4–14.

Stenhouse, L. (1983). *Authority, education and emancipation.* London: Heinemann.

Van Aalsvoort, J. (2004). Activity theory as a tool to address the problem of chemistry's lack of relevance in secondary school chemical education. *International Journal of Science Education, 26*(13), 1635–1651.

Van der Zande, P., Akkerman, S. F., Brekelmans, M., Waarlo, A. J., & Vermunt, J. D. (2012). Expertise for teaching Biology situated in the context of genetic testing. *International Journal of Science Education, 34*(11), 1741–1767.

Van Driel, J. H., Verloop, N., & De Vos, W. (1998). Developing science teachers' pedagogical content knowledge. *Journal of Research in Science Teaching, 35,* 673–695.

Webster, S. (2008). How a Deweyan science education further enables ethics education. *Science & Education, 17,* 903–919.

Wieringa, N., Janssen, F. J. J. M., & Van Driel, J. H. (2011). Biology teachers designing context-based lessons for their classroom practice—The importance of rules of thumb. *International Journal of Science Education, 33*(17), 2437–2462.

Wilmott, C., & Bryantt, J. (2005). Engaging with the ethical implications of science. In *Proceedings of the Science Learning and Teaching Conference* (pp. 85–89).

Wilson, S. M., Shulman, L. S., & Richert, E. R. (1987). "150 different ways" of knowing: Representations of knowledge in teaching. In J. Calderhead (Ed.), *Exploring teachers' thinking*. New York: Taylor & Francis.

Wishart, J., Green, D., Joubert, M., & Triggs, P. (2011). Discussing ethical issues in school science: An investigation into the opportunities to practice and develop arguments offered by online and face-to-face discussions. *International Journal of Science Education, 1*(1), 47–69.

Zeidler, D. L., & Sadler, T. D. (2008). Social and ethical issues in science education: A prelude to action. *Science & Education, 17,* 799–803.

Zeidler, D., Sadler, T., Simmons, M., & Howes, E. (2005). Beyond STS: A research-based framework for socioscientific issues education. *Science Education, 89,* 357–377.

Part II
Case Narratives from India

Chapter 5
Establishing Linkages with Humanistic and Liberal Goals of Science—Case Narrative of a High School Teacher

Abstract The present chapter portrays a case study of a High School Biology teacher named 'Arunima' in order to understand the different facets of her knowledge in the area of 'Ethical Issues in Biological Sciences'. The different knowledge domains that could be studied here include her knowledge about teaching, learning, pedagogy, curriculum, students' misconceptions, assessment, views on Nature of Science (NOS), etc., in the context of ethical issues in Biological Sciences. Teaching and learning about ethical issues in Biological Sciences includes many areas and topics where a need is felt to think beyond the positivist view of Science and incorporate humanistic and liberal ideals for taking the 'right' decisions. It can be clearly seen here that science teaching can no more be an objective affair rather includes many areas where one's own subjectivity and reasoning is reckoned and gets projected. The present case study will provide an insight into the categories of teachers' knowledge that are needed for addressing the ethical issues in a more humanistic way in the Science classroom.

Keywords High school · Biology teacher · Ethical issues in biological sciences · Humanistic approach · Pedagogy · Assessment

5.1 Introduction to the Case

Arunima teaches Biology in one of the reputed schools of Delhi. She teaches Biology to the students of K-12 grade level and holds a teaching experience of around 15 years at higher secondary level. She is postgraduate in Microbiology and possesses a bachelor's degree in Education. Her most favoured topics include Genetics and Biotechnology. She feels that classroom should be an open platform for discussion, freedom of expression as well as clarification of concepts $_{(\mathrm{KTS})}$.[1] She

[1]The present chapter involves the use of certain codes, e.g. KTS, OST, KNOS that refer to teacher's knowledge domains (for detailed reference, see Appendix F).

A. Saxena, *Ethics in Science*, https://doi.org/10.1007/978-981-32-9009-9_5

prefers teaching by way of lecture interspersed with discussion, student interactions as well as audiovisual methods[2] $_{(KTS)}$.

The data collected for the case study was from multiple sources as is prerequisite for a case study type of design. The different sources of data included using surveys and questionnaires, classroom observations, in-depth interviews (both pre-observation and post-observation) and reflective exercises. The data that emerged on a particular ethical issue was narrative in nature and carried implicit ethical concerns. Narrative inquiry as a method was used to delve deeper into the narratives of the teachers and draw interpretations related to the teaching of ethical issues in Biology. An attempt has been made to document the classroom discourses of different cases, wherein an interpretive approach to classroom discourse analysis has been adopted. The data from these different sources was corroborated and finally coded based on the knowledge categories (refer to Appendix F).

5.2 Orientation About Teaching Biology at Senior Secondary Stage

The present teacher feels that studying Biology at senior secondary stage helps in gaining deeper insights into the living world around us, into the environment and its mysteries, and into our own body. She feels that the subject helps in preparing the learners to deal with the daily-life issues in a much more scientific and logical manner $_{(OST)}$. Besides, the subject pertains to the living world or the life sciences, which makes it less abstract and more interesting to teach as well as learn. Some of the major objectives according to her that science education at this level aims to fulfil include development of scientific attitude and scientific temper in solving problems, generating environmental awareness and environmental sensitivity $_{(OST)}$.

5.3 Understanding About Nature of Science (NOS)

The present teacher holds a fairly good understanding about NOS, and holds a positive opinion about the role of history of Science in developing a scientific temper and reasoning, which was evident from her response,

> Definitely, we always learn from history, scientific laws and theories proposed, accepted and rejected helps in development of scientific temper, logical reasoning, etc. $_{(KNOS)}$

She is able to draw interlinkages with the humanistic and liberal goals of Science, by way of certain broader goals of science education, which according to her are, to develop problem-solving ability, to lead scientific research towards

[2]Source: Survey Proforma (Appendix C).

sustainable development, with focus on scientific discoveries yet living in harmony with nature $_{(OST)}$. Logical method of reasoning as per her is the only way of reaching sound conclusions, and students can be given various simulated situations wherein they can apply their ability to reason out and decision-making $_{(KTS)}$. Subjectivity in Sciences which comes from varied scientific, philosophical, intellectual, religious, technological, economic and cultural circumstances does have an impact over scientists and their research, and the present teacher provided the following evidence for it:

> That is the reason why human clones have not come into existence due to religious and ethical pressures on technology and have resulted in making scientific researches highly subjective. $_{(KNOS)}$

She is of the opinion that Science is universal in nature, as scientific principles, theories and laws have a wider applicability, and vouches for democratic scientific practices. The scientists in their labs are not given complete freedom for carrying out research in the desired area. Here, again she feels that certain regulatory committees like GEAC (Genetic Engineering Appraisal Committee) need to be in place so as to keep a check on the researches performed and their outcomes $_{(KNOS)}$.

In dealing with a controversial issue in the classroom, the present teacher thinks that one has to give flexibility for the expression of varied student opinions instead of influencing them with our own. However, one should discourage any kind of negativity and should encourage logical reasoning $_{(KTS)}$.

5.4 General Notions About Ethics in Science

The present teacher certainly supports the role and need for ethics in Science and defends it by saying,

> Yes, definitely, because technology is offering us so many advancements, before using them we need to find out the pros & cons, advantages & disadvantages, and whether it is going against ethics? $_{(KET)}$

She even cited some examples wherein the role of ethics is indispensable such as Genetically Modified crops, human cloning, and use of stem cells. According to her, ethical issues have only been sparsely discussed in the curriculum of K-12 Biology and their incidence needs to be enhanced so as to keep the students abreast with the pros and cons of the latest technology[3] $_{(KCur)}$. This can also be a humanistic ideal of Science, which takes into account the societal norms and interests while deciding upon any given scientific research or technology.

[3]Source: Questionnaire.

5.5 Negotiating Ethical Issues in Biological Sciences

The present teacher adopts varied pedagogical approaches and transactional strategies to deal with the ethical issues related to Biological Sciences. This was evident in her classroom teaching as well where she used newspaper clippings, narration of case studies and discussion on such issues as a regular course of her teaching–learning practices in the classroom. In one of the classes observed on 'Amniocentesis', the following discussion was initiated by her (refer to Box 5.1):

Box 5.1 Classroom Discourse on Amniocentesis

Teacher:	Why should we be worried about female foeticide? $_{\text{(KArg+KCon)}}$
Student 1:	Depression in the sex-ratio.
Teacher:	So What?
Student 2:	More risk to female population, balance in society is also affected.
Teacher:	What will happen when this population reaches reproductive stage?
Student 3:	This could lead to a rise in the incidents of STDs (Sexually Transmitted Diseases).
Teacher:	Is it a sign of a healthy society?
Student 4:	No.
Teacher:	Should the Government Ban this Technique of Amniocentesis? $_{\text{(KArg+KCon)}}$
Student 1:	Yes, it should be banned.
Teacher:	Should the couples have the right to decide whether to have a girl child or a boy child? $_{\text{(KArg)}}$
Student 2:	Why boys only?
Teacher:	More of rural population than urban is misusing this technology, especially in parts of Punjab and Haryana, despite of being educated and knowing that females today are doing even better than males in almost every sphere $_{\text{(SMK)}}$. Sex-determination is punishable by law, but still it does not deter the people from practicing it (deontological) (Here Arunima referred to a column written in a daily newspaper on Female foeticide and asked the students to read it- HT May 07, 2012) $_{\text{(KTS)}}$.

The above classroom excerpt makes evident the misuse of the technology of 'Amniocentesis' resulting in declining sex-ratio that directly impacts the normal functioning of the society and disturbs the nature's balance. Here, there is a clash between the scientific and humanist ideal of Science. The scientific research treats amniocentesis as a useful technology to detect any kind of foetal disorders while the society misuses the technology to detect the gender of the foetus leading to female foeticides, which is against the humanistic ideal. In order to clarify the concepts related to the technique of Amniocentesis, the teacher developed the following blackboard summary. $_{\text{(KTS)}}$

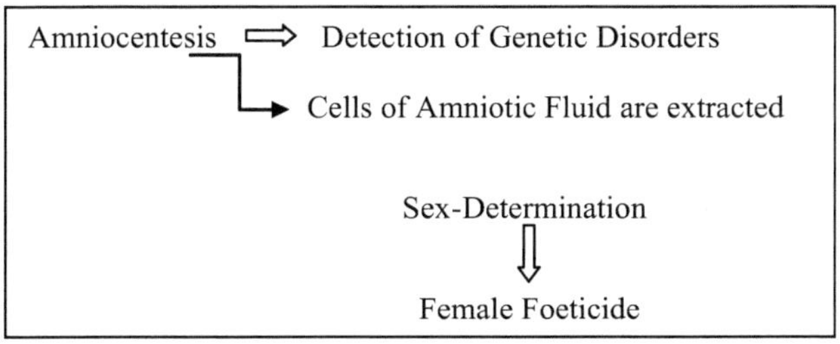

Although the concept and technique of amniocentesis are clear to the students, yet there are many debates about the applicability and accessibility of this particular technology. In another topic on family planning, the teacher discussed in the class about the burgeoning population size and the present situation of the country, which is not equipped to support it due to so many limiting factors, in terms of material resources, job opportunities, education and other medical facilities required for basic human survival. Therefore, a need for controlling the population and practicing family planning arises. Here, various schemes promoted by the government on family planning, awareness programmes, distribution of free contraceptives, free vasectomy and tubectomy, making laws against child marriages, etc., $_{\text{(SMK)}}$ were also being discussed.

She raised a question, "Should Families be planned?" $_{\text{(KArg)}}$ This becomes an ethical issue and is highly debatable as here some religions that strictly prohibit the act of planning the family size stand in opposition to the country's own well-being and the well-being of its citizens that can only be promoted if the population size is controlled $_{\text{(KCon)}}$. Thereafter, a discussion on various methods of contraception such as natural, barrier, chemical and surgical methods were being done in detail. For instance, natural methods are not completely safe, chemical methods such as pills, may have side effects on health, and play havoc with the hormonal balance from prolonged intake $_{\text{(SMK)}}$.

Another topic of ethical relevance is **Contraceptives**, wherein the focus was on surgical methods of contraception that include Vasectomy and Tubectomy, but the students were using the two methods interchangeably. To avoid the confusion between the two terms, the present teacher explained, "Vasectomy pertains to the ligation of Vasa deferentia from where it derives its name, whereas tubectomy refers to the ligation of fallopian tubes." $_{\text{(SMK)}}$

Thus, the teacher kept the discussion focussed on certain technical aspects related to mechanism and use of contraceptives (refer to Box 5.2).

Box 5.2 Classroom Discourse on Contraceptive Methods

Student 1:	Does the menstrual cycle take place after tubectomy?
Teacher:	Yes, the menstrual cycle occurs as the hormone estrogen released by the ovarian follicles enters into the blood stream directly and causes the uterine wall thickening, and hence menstruation takes place. $_{\text{(SMK)}}$
Student 2:	What happens to the ovum released by the ovary?
Teacher:	It gets absorbed by the body and degenerates after some time.
Student 3:	How does Cu-T prevent conception?
Student 4:	It kills the sperms.
Student 5:	It reduces sperm motility.
Teacher:	Copper ions act as irritants and do not allow implantation of embryos in the uterus.
Student 6:	Why do Copper ions have such an effect on the uterus?
Teacher:	They create a non-self like condition in the uterus.
Student 7:	Why Copper ions and not Aluminium ions?
Teacher:	Because Copper is not harmful.

However, she seemed to be less bothered about certain socio-ethical aspects related to the use of such contraceptives, for instance, she could have initiated discussion on some of the following ethical issues associated with the use and working of Contraceptives:

- Do these surgical methods of contraception pose any side effect on the health of a particular individual?
- Why go for surgical methods if alternatives are available?
- Is Cu-T Abortifacient?
- Is taking contraceptive pills a safer option to avert pregnancy?
- What could be some of the side-effects from the repeated usage of these pills (especially, 'i-pills' or emergency contraceptive pills)?
- Is the awareness about all such contraceptive devices leading to better planned families or have increased the incidence of teenage pregnancies and promiscuity?

A discussion about all such issues could have raised an ethical understanding about these contraceptive methods and in highlighting their appropriate usage. Some of the students' difficulties were evident from the above discussion, such as the mechanism involved in surgical methods of contraception, as well as the mechanism of working of Cu-T, and impact of Copper ions on the uterus. These were appropriately addressed by the teacher owing to her sound subject matter knowledge $_{\text{(SMK)}}$.

Another topic was on **'Medical Termination of Pregnancy (MTP)'**, where the present teacher began the discussion by deriving the definition of 'reproductive health' from the students as "a state of total well-being of an individual with respect to reproduction which involves the complete emotional, physical, behavioural and social well-being" $_{(SMK)}$. Then, she referred to a few cases of sexual abuse and its consequences on the health and well-being of the female. The following classroom discourse was initiated by her (refer to Box 5.3):

Box 5.3 Classroom Discourse on Medical Termination of Pregnancy (MTP)

Teacher: "	What happens when a female conceives due to abuse?" $_{(KArg)}$
Student 1:	No response.
Teacher:	The only option available here is Abortion or MTP. It was only in the year 1971 that abortions were legalised in our country (SMK). Why so late? $_{(KCon)}$
Student 2:	Because the technique was being misused for female foeticide.
Teacher:	Around 40-50 million abortions take place every year. Is it good or bad to legalise abortions?
Students:	[Some say good and some bad, but not in a state to decide]
Teacher:	Abortion should be legalised as there could be many instances when the couple doesn't want a child, for instance, in the case of unplanned pregnancies, any kind of abnormality in the fetus, etc. however sex-determination is illegal and should be strictly banned $_{(TB)}$.

Thus, in the above classroom discourse, the teacher clarifies that MTP if carried out under some conditions, such as, in the case of sexual abuse or foetal abnormality could be justified but if done just for sex-selection, it is unethical. Another area that is beset with many ethical issues is Assisted Reproductive Technology (ARTs), of which In Vitro Fertilization (IVF) has generated many debates and controversies. The teacher initiated following discussion in her class (refer to Box 5.4).

Box 5.4 Classroom Discourse on In Vitro Fertilization

Teacher:	Have you heard about a woman given birth to triplets? (The teacher then reads from a newspaper clipping- "Quintuplets born after infertility treatment die", with this the discussion in the class started about the utility and risks associated with the infertility treatment.) $_{(KTS+KCon)}$ Teacher asked the students- "Was such an infertility treatment good?" $_{(KArg)}$
Student:	It is not always necessary that the infants die.

Teacher: Since the Infertility rate is increasing these days due to pollution and stressful lifestyles, which has given rise to these technologies. Thereafter the teacher narrated a real-life story of an infertile couple,
"A couple was diagnosed with the infertility problem and were advised medication by a doctor in order to conceive a child, however the couple were from a relatively low socio-economic background and could not afford the expensive medication and treatment thereafter. As the doctor came to know about it, she herself visited the couple at their home and provided them financial support for facilitating the treatment" (KTS+KCon).

A point was raised about concealing the identity of the donor (whether an egg or a sperm donor) for preventing any ethical or social issues, but was not discussed in detail. An argument that was presented by the teacher here was, the procedure of in vitro fertilization involves the formation of many embryos as many eggs are being simultaneously fertilized, but only one can be implanted into the uterus, then what happens to the rest? (KArg) As explained by the teacher, they are cryopreserved for further use in case the implanted embryo does not survive. (SMK)

Thus, from the above classroom episode, it is evident that the topics such as Amniocentesis, MTP contraceptives and ARTs are beset with many ethical issues, which were implicitly surfaced by the teacher in the class. There are however some grave ethical issues pertaining to estranged relationships between surrogate mother and child, real mother and child developed through ART, treatment given to such children (test tube babies) in the society, are they being accepted as normal human beings or included under special category, what could be the impact of such altered attitude on the self esteem of such children? Such questions are ethical in character and thus require a mode of ethical inquiry and argumentation to resolve them.

Biotechnology and its applications is another area spearheading many scientific researches and have a number of ethical concerns attached to it. In order to deal with the topic, the present teacher chose a newspaper clipping on "Birth Waste to Heal wounds faster" and another on "The Great Cotton Rush" (KTS). These were discussed with the students to bring out the issues and concerns attached with the technique of In Vitro Fertilization (IVF) and Genetically Modified (GM) crops. (SMK) An excerpt from the classroom discussion can be taken as follows (refer to Box 5.5).

Box 5.5 Classroom Discourse on Genetically Modified (GM) Crops

Teacher: What do we mean by the great cotton rush and to whose benefit it might be? (KCon)
Student 1: To the farmers who are growing cotton.
Teacher: Earlier the cotton yield was drastically affected and destroyed by an insect pest called as 'Boll worm' that infested the cotton fields. What solution was proposed by the scientists? (SMK)
Student 2: Use of Biotechnology to develop pest-resistant crop-plants.
Teacher: It is a win-win situation for farmers as once they buy Bt-cotton seeds from the market, later they can grow their own Bt-seeds. Now, is it possible for a farmer who does not have any knowledge about Bt-cotton to grow his own Bt-seeds?
Student 3: Yes, the farmers can grow their own Bt-seeds with each yield by the process of self-pollination.
Teacher: Why Bt-cotton and not natural variety?
Student 4: This is because it reduces the cost of production by lesser to negligible use of pesticides.
Teacher: In Orissa, many farmers were committing suicide as they did not have money to invest in pesticides and fertilisers, therefore the only option left to them was to grow Bt crops (KArg+KCon).

The ethical issues attached to the production and consumption of GM crops that were taken up by her included breach of certain religious beliefs, and unprecedented consequences on human health and environment.

Some other important applications of recombinant Biotechnology such as Vitamin A-rich Rice, genetically engineered insulin, gene therapy and treatment of certain genetic disorders were also discussed (SMK).

A Summary of Classroom Processes in Dealing with Ethical Issues in Biological Sciences

S.No.	Classroom processes			
1.	Introduction to the Topic	Beginning with a recapitulation/vignette/newspaper article related to the topic		
2.	Kind of examples/analogies used	From daily life and personal experiences	From the textbook (if any)	From television and other media
3.	Number of ethical arguments raised	3–4 per class		
4.	Type of ethical arguments raised (along with some evidence)	**Deontological** (For example, Should the couples have the right to decide whether to have a girl child or a boy child?) **Utilitarian** (For example, "Quintuplets born after infertility treatment die", with this the discussion in the class started about the utility and risks associated with the infertility treatment.) Teacher asked the students—"Was such an infertility treatment good?") **Consequentialist** (For example, Besides being a pest-resistant variety, Bt Brinjal was stopped from coming into the market. Why? and who opposed it?)		

(continued)

(continued)

S.No.	Classroom processes	
5.	Treatment given to ethical arguments	Listed, discussed
6.	Students' contribution to ethical arguments	Just a few students active in discussion
7.	Classroom time devoted to the teaching of ethical issues	5–10 min
8.	Kind of resolution reached	An ethical dilemma created in students' minds
9.	Major pedagogical approach adopted	Lecture interspersed with discussion
10.	Kind of understanding aimed at by the teachers	Conceptual

5.6 Understanding with Respect to the Ethical Issues in Biological Sciences

It was during in-depth interview sessions as well as post-observation interviews, that teacher's understanding on some of the ethical issues related to Biological Sciences could be gauged. The theme-wise list of topics with associated ethical issues was used for the in-depth interviews with the teacher. The responses given by the teacher were mostly in the form of arguments to support or oppose a particular stance. These have been lifted as it is from her interview transcripts and presented below in a tabular form (refer to Table 5.1). Since the responses had some inherent reliance on an ethical theory or principle, therefore an analysis of responses has been done alongside based upon some major ethical theories (refer to Chap. 1).

From the in-depth interview analysis of the teacher's arguments, it is evident that she is definitely in the favour of technology and biotechnological advances. This was especially evident when she said,

> If we ban the technique then all the benefits that such a technique has to offer are also removed. Then what is the use of banning such a technique that has a good positive effect also. $_{(SMK+TB)}$
>
> Contraceptives HAVE to be used as there is population explosion, which is the root cause of all the problems." $_{(SMK+TB)}$
>
> As it is every month an egg is being wasted during menstruation, even in an ejaculate there are millions of sperms, out of which only one fertilises, so here again there is a wastage of sperms. We have to be somewhat liberal, we cannot think on these grounds, IVF is helping so many couples so if some embryos are being wasted then we have to bear with it, you know every rose has a thorn. $_{(TB)}$

Table 5.1 Analysis of teacher's arguments based on ethical theories

S. No.	Topic	Ethical issues	Teacher's SMK/Opinions/Beliefs/Metaphors	Related ethical theory
1.	Animal experimentation	1.1. Use of animals in research and study 1.2. Pain and suffering to the animal 1.3. Wastage of animals	"It allows in understanding many systems together and at least with rats which are not few in number." (KArg) "Animals should be shared and teachers should see to it that students do not waste animals." (KTS)	Utilitarian
			"Animals are also equal living beings as we are and we have no right to kill them as such. But, for study purposes a student of Biology needs to have an insight as to what the animal looks like from inside." (KArg)	Ethic of care
2.	Amniocentesis	2.1. Sex determination 2.2. Female foeticide 2.3. Craving for the male child	"Sex-determination has already been banned but if we ban the technique then all the benefits that such a technique has to offer are also removed. Then what is the use of banning such a technique that has a good positive effect also." (KArg+TB)	Utilitarian
3.	Contraception	3.1. Religious ban on contraceptives 3.2. Contraceptives can be abortifacient 3.3. Fringing right to birth and right to life 3.4. Conceptus as a potential human being	"Contraceptives HAVE to be used as there is population explosion, which is the root cause of all the problems." (SMK+TB)	Utilitarian
			"Yes, some contraceptives could be abortifacient and they should not be used, for example Cu-T will only prevent implantation, fertilization can still occur." (SMK+TB)	Consequentialist
4.	MTP or abortion	4.1. Right to birth and right to life 4.2. Foetus as a potential adult 4.3. Principle of sentience 4.4. Instances where MTP is ethically appropriate	"We can't say as it depends on the conditions. Sometimes people are left with no option but to abort. For example, in case of single mother." (TB)	Principle of Non-Maleficence

(continued)

Table 5.1 (continued)

S. No.	Topic	Ethical issues	Teacher's SMK/Opinions/Beliefs/Metaphors	Related ethical theory
5.	In vitro fertilization	5.1. Safety of the technique in terms of mother's health 5.2. Wastage of embryos 5.3. Estranged relationships between child–parent, child–sibling, child–society, etc.	"There is no harm in using this technology if it can provide them with a baby; the only thing is nothing should be exploited beyond its limits." (TB) "As it is every month an egg is being wasted during menstruation, even in an ejaculate there are millions of sperms, out of which only one fertilises, so here again there is a wastage of sperms. We have to be somewhat liberal, we cannot think on these grounds, IVF is helping so many couples so if some embryos are being wasted then we have to bear with it, you know every rose has a thorn." (TB)	Utilitarian
6.	GM crops	6.1. Do we really need GM? 6.2. After effects of GM on human health and environment 6.3. Accessibility and means to buy GM crops/seeds 6.4. Is organic food a better option? 6.5. Gene patents and biopiracy	"If we talk about GM Cotton then there is no harm in its cultivation as nobody is going to eat it, but if we talk about Bt-Brinjal, yes even I won't like to eat because you never know what thing it is going to cause." (TB) "If we use Bt-cotton for a long time it may create resistance in boll worms, so the alternative is that is being practiced by people in America who now grow one row of Bt-cotton and one row of normal cotton. Such a practice helps in preventing resistance amongst boll worms." (SMK) "Organic farming is good, but has to be proved on a large scale. The crops available as organic produce are so costly, an ordinary man cannot afford." (TB)	Utilitarian
			"If we don't have patents then the result would be that foreign countries come, use our resources, and issue a patent claiming that it is a product of their origin, which is wrong and India is very late in setting up these patents." (SMK)	Principle of justice

(continued)

Table 5.1 (continued)

S. No.	Topic	Ethical issues	Teacher's SMK/Opinions/Beliefs/Metaphors	Related ethical theory
7.	Genetically Modified Organisms (GMOs)	7.1. Misuse of GMOs as bioweapons 7.2. Pain and suffering experienced by the concerned animal 7.3. "Playing God" and tinkering with the blueprints of life forms 7.4. Intrinsic respect and consideration for other life forms 7.5. "Transgenesis" leading toward breaching of species barrier	"There is no harm if we are exploiting bacteria to provide us with insulin. But, if used as bio weapons, they can be disastrous." (SMK)	Utilitarian
			"GEAC, that agency has to be consulted before creating a GMO. So, that keeps a check on GMOs, but regarding bio weapons and all which are done through illegal processes, a check is difficult to sustain and this is a sheer misuse of technology." (SMK) "We are creating GMOs for drugs and vaccine research and yes the animals do have to bear through all the pain and suffering. But, then there are always two sides of the coin." (SMK) "The research has to be done but limited and restricted and done only in those cases where it is very urgent with minimal wastage and inhuman exploitation of animals." (SMK)	Principle of nonmaleficence
8.	Human genomic project	8.1. Principle of privacy and confidentiality of genetic information 8.2. Playing with nature's diversity and gene pool 8.3. Issue of risking human life 8.4. Issue of informed consent 8.5. Gene patent	"I don't think that we are anywhere loosing the privacy. And that will help us detect diseases easily, and can treat them in a better manner. Nowadays, doctors do give drugs but many-a-times it is a hit and trial method that they adopt. But, if our genome is available then specific drugs can be given." (SMK+TB) (Genetically engineered babies): "We can't predict with surety and we may land up with something highly abnormal. So, this thing should not be given in the hands of humans to play with." (SMK+TB)	Utilitarian
			"Genetic screening of an individual's genome should be done only in teachers where it is very much needed such as criminal teachers, or identity issues, otherwise, there is no use of knowing human genome unless for detection of genetic diseases." (SMK+TB)	Consequentialist

Source In-depth interviews with the case

5.7 Sensitivity Towards Ethical Issues

Most of the arguments proposed by the present teacher are 'Utilitarian' in their character as she keeps in mind the ultimate benefit to mankind such as in alleviating the pain and suffering, treatment of certain disorders and bringing happiness in people's lives. Some elements of teacher's emotional ecology (Zembylas, 2007) were recorded during the in-depth interviews and classroom teaching. These have

Table 5.2 Arunima's emotional metaphors

Planes	Emotional metaphors
Individual	**Animal experimentation** I remember the experiment of frog where it was not killed but was pithed by hitting the head on the slab; it is "AWFUL AWFUL (pain)". That should be "totally banned", no ethics, nobody is ever concerned what is happening. That is "so inhuman…." **Amniocentesis** Sex determination is illegal and should be "strictly banned" **IVF** "We have seen so many couples facing this problem of infertility which is all due to lifestyle changes and infertility problems." "If some embryos are being wasted then we have to bear with it, you know 'every rose has a thorn'." **GM crops** It is a "win-win" situation for farmers as once they buy Bt cotton seeds from the market, later they can grow their own Bt seeds Due to some religious reasons as they feel that these crops have some "animal genes" and hence not considered under a vegetarian diet It may not cause any harm to us in the near future but maybe in the far future, we do not know what are the repercussions of having a product in which a bacterial gene has been incorporated **Biopatents** YES, definitely, I think there's a "dire" need **GMOs** If used as bioweapons, they can be "disastrous" Yes the animals do have to bear through all the "pain" and "suffering". But, then there are always "two sides of the coin" **Teaching of ethical issues** A teacher when dealing with such issues has to deal with both the "pros and cons side of technology" otherwise the students would never come to know that there could be this "other side of technology" also which can be harmful There is more of "mugging up", wherein "knowledge is given" and child has to keep on mugging up. And there is hardly any scope for critical thinking. What has been discovered is being told to the child, and the child has to learn it
Relational	We need to make them "environment conscious" If a child is studying Biology, he should be aware of the controversies related to the technology, and "where to set the limits"
Sociopolitical	There is corruption everywhere

also been analysed under planes of ecology and have been listed as 'emotional metaphors'[4] (marked in paranthesis) in Table 5.2.

5.8 Analysis of Arguments

In order to better ascertain the quality of ethical arguments raised by the teacher, these have been analysed using Toulmin's method of argument analysis (Toulmin, 1958, 2003). Each of the argument has been dissected out based upon the five different components, viz., data, warrant, qualifier, rebuttal and claím (refer to Appendix H). From the analysis of the teacher's arguments, it is evident that most of the arguments have all the six components which make them qualitatively sound as well as complete (refer to Table 5.3).

The analysis of arguments reveals an average score of 2.4 and is low as compared to other teachers. This can be attributed to various factors such as knowledge and awareness about the issues of ethical concern, sensitivity towards ethical issues, lack of adequate evidence to support a claim, weak counterarguments and rebuttals.

5.9 Perception About the Curriculum at Senior Secondary Stage

The present teacher feels that the curriculum at senior secondary level should not be integrated as it dilutes the content and the subject's essence is lost, but she does feel that certain topics need an integrated approach, for instance, the Biomolecules chapter in class XI, where integration between the Biology and Chemistry concepts is required. The topic according to her has been given a superficial treatment and hence makes it vague for the students to understand. She feels that contentwise, the textbook needs a revision and renewal.

> I think it is high time that the book needs a renewal, NCF came in 2005 and now it is 2013, for a subject like science, if we are not able to update the book even in 7 years, so many researches have come about, so why not to put those latest research content into the book, which makes it more interesting. Stem-cells they have just explained the term, why not give them more articles, in text content about where stem cell therapy has been used, more examples, Biotechnology has so much advanced. $_{\text{(KCur)}}$

[4]Emotional Metaphors indicate how someone feels about events, objects or people in his/her life. Researches have also shown that how one's professional knowledge in teaching may be identified and analyzed through the use of one's metaphorical language. Metaphorical language is said to be an extraordinarily powerful tool through which teachers express the meaning of their experiences in teaching and much of their pedagogical knowledge which teachers hold (e.g., see Briscoe, 1991; Clandinin, 1986).

Table 5.3 Argument analysis based on Toulmin's method

S. No.	Argument topic	Data (a phrase to explain)	Warrant	Qualifier	Backing	Rebuttal	Claim
1.	Animal experimentation (Level = 3)	It allows in understanding many systems together and at least with rats which are not few in number	Animals should be shared and teachers should see to it that students do not waste animals	Students do not waste animals	Ethic of care & concern Animal rights	Animals are also equal living beings as we are and we have no right to kill them as such	But for study purposes, a student of Biology needs to have an insight as to what the animal looks like from inside
2.	Amniocentesis (Level = 2)	If we ban the technique then all the benefits that such a technique has to offer are also removed More of the rural population than urban is misusing this technology, especially in parts of Punjab and Haryana Sex determination is punishable by law, but still it does not deter the people from practicing it			Utilitarian theory and principle of beneficence		What is the use of banning such a technique that has a good positive effect also

(continued)

Table 5.3 (continued)

S. No.	Argument topic	Data (a phrase to explain)	Warrant	Qualifier	Backing	Rebuttal	Claim
3.	Contraception (Level = 3)	Burgeoning population size and the present situation of the country So many limiting factors, in terms of material resources, job opportunities, education and other medical facilities required for basic human survival	There is population explosion, which is the root cause of all the problems		Consequentialist	Yes, some contraceptives could be abortifacient and they should not be used, for example, Cu-T will only prevent implantation, fertilization can still occur	Contraceptives have to be used
4.	In vitro fertilization (Level = 2)	IVF is helping so many couples so if some embryos are being wasted, then we have to bear with it, you know every rose has a thorn.	As it is every month, an egg is being wasted during menstruation, even in an ejaculate, there are millions of sperms, out of which only one fertilizes, so here again, there is wastage of sperms. We have to be somewhat liberal, we cannot think on these grounds	The only thing is nothing should be exploited beyond its limits	Utilitarian		There is no harm in using this technology if it can provide them with a baby

(continued)

Table 5.3 (continued)

S. No.	Argument topic	Data (a phrase to explain)	Warrant	Qualifier	Backing	Rebuttal	Claim
5.	GM Crops (Level = 3)	It is a win-win situation for farmers as once they buy Bt cotton seeds from the market, later they can grow their own Bt seeds	Alternative is that it is being practiced by people in America who now grow one row of Bt cotton and one row of normal cotton. Such a practice helps in preventing resistance among bollworms	If we talk about Bt Brinjal, yes even I do not like to eat because you never know what thing it is going to cause	Principle of beneficence and nonmaleficence	If we use Bt cotton for a long time, it may create resistance in bollworms	If we talk about GM cotton, then there is no harm in its cultivation as nobody is going to eat it
6.	GMOs (Level = 3)	We are creating GMOs for drugs and vaccine research	There are always two sides of the coin	If used as bioweapons, they can be disastrous. Research has to be done but limited and restricted and done only in those teachers where it is very urgent with minimal wastage and inhuman exploitation of animals	GEAC, that agency has to be consulted before creating a GMO. So that it keeps a check on GMOs	The animals do have to bear through all the pain and suffering	There is no harm if we are exploiting bacteria to provide us with insulin
7.	HGP (Level = 1)	Nowadays, doctors do give drugs but many a times, it is a hit and trial method that they adopt	I don't think that we are anywhere loosing privacy. And that will help us detect diseases easily, and can treat them in a better manner		Principle of beneficence utilitarian		If our genome is available then specific drugs can be given

The ethical validity according to her has been the most neglected part of the curriculum as the treatment given to ethical issues in the textbook or in the curriculum is very superficial. However, according to her, it depends upon the individual teacher, how she takes care of these ethical issues in the classroom while teaching. She holds the idea of technology that is in harmony with nature, which means technology that poses the least harm to the people and their environment. Here, the teacher takes recourse to 'scientific humanism' that gives preponderance to understanding and law of beneficence that does the least harm to the human race (Shermer, 2019). The role of a teacher as propounded by her has been summarized in the following statements made by her during the interview:

> Any technology which we are teaching to the students we should take into account not only its benefits but also its side-effects on nature, on environment, on soil, on humans. So, every time we do this we are sensitising the child toward the effects of advancement on nature. Such that later in life when the child grows up and uses whatever technology, then he knows that this harm could be caused by this technology and hence take responsible decisions in life as well become a responsible citizen. A teacher when dealing with such issues HAS TO DEAL with both the pros and cons side of technology otherwise the students would never come to know that there could be this other side of technology also which can be harmful $_{(KTS)}$. And they will not become environment conscious and we need to make them environment conscious.

According to her, the curriculum of Biology makes the student memorize the content, rather than initiating them into critical thinking and problem-solving process $_{(KCur)}$. Discussing ethical issues at this level is important, as they open up students' minds, motivates them to think, reason out, reflect, become responsible consumers of technology and also helps in character education (Zeidler & Nichols, 2009). Inclusion of ethical issues within the curriculum also makes the lesson more interesting $_{(KCur)}$. She places the ethical issues within the chapter rather than sidelining them or leaving them for the end. This was evident when she said,

> Ethical issues should be placed within the chapter, whenever I am teaching a concept; I wish to discuss the issues then and there. I need not tell the students that these are ethical issues, but enable them in reasoning out and helping them reach a conclusion. $_{(KCur)}$

In this case, the teacher tried to bring in the concept of Critical Consciousness and raising conscientious souls (Freire, 1972) who can think critically on the social, political, legal and scientific issues surrounding the issue of GM.

5.10 Knowledge About Planning

In order to find out as to what goes on in the teacher's mind while planning for her classes, a planning sheet was given to the case and she was asked to fill it up. The responses gathered from the sheet have been tabulated as follows.

Topic 1 (any ethical issue): ARTs (Assisted Reproductive Techniques)
Class Time: 35 min

Planning (Activity/ Experiment/ Demonstration/ Discussion/etc.)	**Mode of execution** (Process involved along with time required)	**Desired outcome** (Intended objectives)	**Assessment**
Newspaper articles	Discussion on the technique of IVF by incorporating some advertisements, reading newspaper items, narrating some cases of IVF babies and their parents, thus situating in the context. (KCon)	To make the learners socially aware To guide them towards the right kind of decision-making through discussion	Home
assignments			

Topic 2 (any topic other than ethical issues): Ecology (working of a sewage treatment plant)
Class Time: 35 min

Planning (Activity/ Experiment/ Demonstration/ Discussion/etc.)	**Mode of Execution** (Process involved along with time required)	**Desired outcome** (Intended objectives)	**Assessment**
Written a letter to the Delhi Jal Board, one month in advance for a field visit	Visit the Delhi Jal Board, sewage treatment plant, Keshopur, New Delhi, India Studied and taught about primary and secondary treatment of sewage, filtration, sedimentation, etc.	Students were able to practically visualize how the sewage is treated Students were also able to differentiate between the policy perspective and its praxis	Feedback

From the planning sheet, it is evident that planning of topic 2 (working of a sewage treatment plant) is more elaborate and systematic as compared to that for topic 1 (ARTs). This is in terms of planning the teaching–learning strategy, objectives framed, mode of execution and assessment.

5.11 Discussion

The data collected from classroom discourse, interviews and questionnaires depicts a picture of PCK that is more dependent upon the Subject Matter Knowledge $_{(SMK)}$ and Teacher's Beliefs $_{(TB)}$ (refer to Table 5.4 and Fig. 5.1). At some point, ethical discourse was seen happening in the classrooms, but with very little participation from students' side, as most of the times, it was the teacher who acquired the centre stage and guided the discussion. Teacher's subject matter knowledge generally comes from the textbook of Biology which is NCERT textbook of K-12 level I, and

Table 5.4 Teacher's knowledge components constituting PCK

S.No.	Teacher's knowledge component (Code)	Evidence (in class-room interaction, discussion, interview, questionnaires, lesson plans, reflective fastwrites)	Frequency
1.	Subject Matter Knowledge (SMK)	"Vasectomy pertains to the ligation of Vasa deferentia from where it derives its name, whereas tubectomy refers to the ligation of fallopian tubes." "Some other important applications of recombinant biotechnology such as Vit-A rich Rice, genetically engineered insulin, gene therapy and treatment of certain genetic disorders."	25
2.	Teacher's Beliefs (TB)	"If we talk about GM Cotton then there is no harm in its cultivation as nobody is going to eat it, but if we talk about Bt-Brinjal, yes even I won't like to eat because you never know what thing it is going to cause." "There is no harm in using this technology if it can provide them with a baby; the only thing is nothing should be exploited beyond its limits."	14
3.	Knowledge of Transactional Strategies (KTS)	The teacher even quoted the example from the recent series 'Satyamev Jayate', in its recent segment on 'Child Abuse', which proves that our society is a reproductively unhealthy society. "A teacher when dealing with such issues HAS TO deal with both the pros and cons side of technology otherwise the students would never come to know that there could be this other side of technology also which can be harmful."	12
4.	Knowledge of the Curriculum (KCur)	The curriculum of Biology, the teacher feels makes the student mug up the content, rather than enthusing them into critical thinking and problem-solving "Ethical issues should be placed within the chapter, whenever I am teaching a concept; I wish to discuss the issues then and there. I need not tell the students that these are ethical issues, but enable them in reasoning out and helping them reach a conclusion."	4

(continued)

Table 5.4 (continued)

S.No.	Teacher's knowledge component (Code)	Evidence (in class-room interaction, discussion, interview, questionnaires, lesson plans, reflective fastwrites)	Frequency
5.	Knowledge of Students' Understanding (KSU)		0
6.	Knowledge of Context (KCon)	Narrating some teachers of IVF babies and their parents, thus situating in the context Should the government ban this technique of amniocentesis?	10
7.	Knowledge of Assessment (KA)		0
8.	Knowledge of ethical theories/ Role of ethics (in general and in Science) (KET)	"I support research to be done with caution, environment sensitivity and human benefit in mind." (Utilitarian)	2
9.	Orientation towards Science Teaching (OST)	Certain broader goals of science education, which according to her are, to develop problem-solving ability, to lead scientific research towards sustainable development, with focus on scientific discoveries yet living in harmony with nature	3
10.	Knowledge of ethical Argumentation (raising arguments/ initiating argumentation/resolving) (KArg)	"Some people even have reservation against GM crops, due to some religious reasons as they feel that these crops have some animal genes and hence not considered under a vegetarian diet." "Should Families be planned?" "Sex-determination has already been banned but if we ban the technique then all the benefits that such a technique has to offer are also removed. Then what is the use of banning such a technique that has a good positive effect also."	13
11.	Knowledge about Nature of Science (KNOS)	"Definitely, we always learn from history, scientific laws and theories proposed, accepted and rejected helps in development of scientific temper, logical reasoning, etc."	18
		PCK_{TOTAL}	**101**

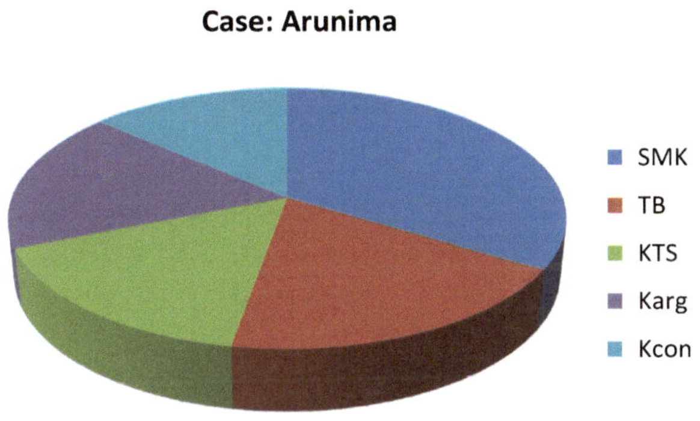

Fig. 5.1 Graphical representation of components of PCK

some newspaper articles. The textbook also reflects the ideology of the sociopolitical system of the state (Fensham, 1992) and adopt the status quo (Hurd, 1989; Roberts, 1995) rather than aligning the curricula with the recent research, evidence and approaches, context and life of the learners. As a result of this, teachers get very few opportunities to discuss ethical issues at greater length in the classroom. The transactional strategies with respect to dealing with ethical issues in Biology most commonly adopted by the present teacher included lecture method; didactic approach (which was mostly instructional and less interactive); reading out newspaper articles, mainly the headlines and generating some discussion in the class. However, very few students in her class were participating or coming up with their viewpoints or raising questions, most of them were passive learners. In classroom discussions, as well as in the interview sessions, there were hardly any instances which could reflect her understanding about the students' difficulties in understanding these ethical issues, their alternative conceptions about ethical issues in Science. A fair number of ethical arguments were raised by her in the classroom and during the interview sessions that indicate her knowledge about the integration of ethics in science and technology. While presenting any issue of science and technology, she tends to give pro as well as anti arguments, such as evident from the following statements,

> There is no harm if we are exploiting bacteria to provide us with insulin. But, if used as bio weapons, they can be disastrous.

> Animals are also equal living beings as we are and we have no right to kill them as such. But, for study purposes a student of Biology needs to have an insight as to what the animal looks like from inside.

The ethical theory that she relied upon for raising arguments with respect to ethical issues happened to be Utilitarian and Consequentialist theory. Some of the arguments made by the teacher indicate a pro-technology approach while dealing with ethical issues, such as

> IVF is helping so many couples so if some embryos are being wasted then we have to bear with it, you know every rose has a thorn.

> We are creating GMOs for drugs and vaccine research and yes the animals do have to bear through all the pain and suffering. But, then there are always two sides of the coin.

Thus, she seems to have a limited understanding about the ethical issues underlying the technology of IVF and GMOs, or simply considers only the benefits that the technology has to offer. There was no specific assessment strategy planned by the teacher in this regard, which again reinforces the notion that these ethical issues are being given secondary treatment within the curriculum as well as in teaching learning process.

5.12 Conclusion

The present teacher has an ample subject matter knowledge, but has a limited understanding about the importance and scope of ethical issues in Biological Sciences, though she is able to address quite a few of them appropriately in the classroom. The reason can be attributed to (as depicted in Fig. 5.1) her lack of knowledge about argumentation and varied perspectives associated with the ethical issues, limited transactional strategies to deal with ethical issues in the classroom, lack of students' understanding as well as lack of knowledge of the context (which includes the sociopolitical–economic and ethical). In fact, the curriculum that she is given to follow also poses a hindrance in discussing and elaborating on these issues. In order to integrate the humanistic ideals in Science curriculum, the entire curricular structure needs a revamp towards a humanistic ideal, wherein these ethical issues should feature as an important area of concern and deliberation rather than being sidelined. There has been a vehement criticism of the traditional Science curricula at least in the West, which emphasises technical know-how and presents the content in a manner which is socially sterile, impersonal, frustrating, intellectually boring and dismissive of students' life-worlds (Hurd, 1989; Lee & Roth, 2002; Reiss, 2000). It was in the late twentieth century that a change in the orientation towards a humanistic ideal grew among the researchers and curriculum makers, especially in the U.S, U.K, Australia and the Netherlands (Aikenhead, 2007). This included the philosophical and sociological perspective in Science and the wave of STS (Science, Technology & Society) approach in curriculum which integrated the socio-scientific perspective to Science making it a more responsible endeavour both for the scientists as well as citizens (Aikenhead, 1980; Bybee, 1993; Cross & Price, 1992, 2002; Kortland, 2001). A major conception towards humanistic Science curricula emerged during post-World War II and emphasised the poststructuralist analysis of Science itself, which is chiefly marked by Kuhn's 'Structure of Scientific Revolutions' published in 1962. such a conception challenged the positivist and realist nature of traditional Science and scientific courses (Abd-El-Khalick and Lederman, 2000). Humanistic perpective in

Science represents an approach to make Science relevant in solving real problems arising from scientific applications in the society. For instance, whether a particular industrial project be sanctioned even if the scientific evidence supports it, but how will it affect the society and the surrounding environment is what determines decision-making in a humanistic Science curricula. But in India, there are fewer studies that espouse a need for the humanistic and STS approach to the curriculum and also a lack of teacher training and education in the same respect has further accentuated the gap in understanding. It is suggested that further research in this domain of Socio-Scientific issues and ethical inquiry can spearhead the movement to promote the humanistic ideal in science education rather than blindly adopting the pro-technology approach.

References

Abd-El-Khalick, F., & Lederman, N.G. (2000). Improving science teachers' conceptions of nature of science: A critical review of the literature. *International Journal of Science Education, 22*, 665–701.

Aikenhead, G. S. (1980). *Science in social issues: Implications for teaching*. Ottawa, Canada: Science Council of Canada.

Aikenhead, G. S. (2007). Humanistic perspectives in the science curriculum. In S.K. Abell & N.G. Lederman (Eds.), The handbook of research on science education (pp. 881–910). New Jersey: Lawrence Erlbaum Associates, Inc.

Briscoe, C. (1991). The dynamic interactions among beliefs, role metaphors, and teaching practices: A case study of teacher change. *Science Education 75*(2), 185–199.

Bybee, R. W. (1993). *Reforming science education*. New York: Teachers College Press.

Clandinin, D. J. (1986). Classroom practice: Teachers' images in action. London: Falmer Press.

Cross, R. T., & Price, R. F. (1992). *Teaching science for social responsibility*. Sydney: St. Louis Press.

Cross, R. T., & Price, R. F. (2002). Teaching controversial science for social responsibility: The case of food production. In W.-M. Roth & J. Désautels (Eds.), *Science education as/for sociopolitical action* (pp. 99–123). New York: Peter Lang.

Fensham, P. J. (1992). Science and technology. In P. W. Jackson (Ed.), *Handbook of research on curriculum* (pp. 789–829). New York: Macmillan Publishing Co.

Freire, P. (1972). *Pedagogy of the oppressed*. U.S.A: Penguin Books.

Hurd, P. (1989). Science education and the nation's economy. In A. B. Champagne, B. E. Lovitts & B. J. Calinger (Eds.), *Scientific literacy* (pp. 15–40). Washington DC: AAAS.

Kortland, J. (2001). *A problem posing approach to teaching decision making about the waste issue*. Utrecht: Cdβ Press.

Lee, S., & Roth, W. M. (2002). Learning science in the community. In W.M. Roth & J. Désautels (Eds.), *Science education as/for sociopolitical action* (pp. 37–66). New York: Peter Lang.

Reiss, M. J. (2000). Understanding science lessons: Five years of science teaching. Milton Keynes: Open University Press.

Roberts, D. A. (1995). Junior high school science transformed: Analysing a science curriculum policy change. *International Journal of Science Education, 17*, 493–504.

Shermer, M. (2019). The case for scientific humanism. Retrieved March 17, 2019, from https://www.scientificamerican.com/article/the-case-for-scientific-humanism/.

Toulmin, S. (1958). *The uses of argument*. New York: Cambridge University Press.
Toulmin, S. (2003). The uses of argument- an updated edition. U.K: Cambridge University Press.
Zeidler, D. L., & Nichols, B. H. (2009). Socioscientific issues- theory and practice. *Journal of Elementary Science Education*, 49–58.
Zembylas, M. (2007). Emotional ecology: The intersection of emotional knowledge and pedagogical content knowledge in teaching. *Teaching and Teacher Education, 23*(4), 355–367.

Chapter 6
Incorporating Ethics for Decision-Making—Case Narrative of a High School Teacher

Abstract The present chapter presents a case study of a High School Biology teacher named 'Saira' and aims to understand the different facets of her knowledge in the area of 'Ethical Issues in Biological Sciences'. She takes recourse to different ethical theories, such as Virtue theory, Deontological and Utilitarian theory which though implicit are evident in her arguments and approach towards teaching–learning of ethical issues in Biology. Narrative inquiry as a method was used to probe into teacher's understanding and conceptions related to the teaching of Bioethical issues. An attempt has been made to document the classroom discourses between the teacher and her students on different ethical issues, wherein an interpretive approach to classroom discourse analysis has been adopted. Teaching and learning about ethical issues in Biological Sciences involves ethical reasoning and ethical inquiry unlike the scientific method of investigation. The present case study opens a new window to address and view the ethical issues in Science from a subjective viewpoint and facilitate decision-making that is in consonance with the humanistic ideal of Science which is all encompassing and socially justifiable. It will further provide an insight into the categories of teachers' knowledge that are needed for addressing the ethical issues in the classroom.

Keywords High school · Biology teacher · Ethical issues in biological sciences · Pedagogy · Assessment

6.1 Introduction

Saira is a senior Biology teacher at one of the reputed private schools of Delhi. She holds Masters degree both in Science and in Education and has over 30 years of experience in teaching students at the higher secondary level. She teaches Biology at K-12 level and her favourite topics include new developments in the field, latest

A. Saxena, *Ethics in Science*, https://doi.org/10.1007/978-981-32-9009-9_6

advances in medical technology, etc. She has a passion for teaching and likes to involve her students in classroom discussions. She uses charts and newspaper clippings to best explain the concepts.[1]

6.2 Notions About Teaching Biology at Senior Secondary Stage

Saira's basic notion about teaching Biology is to create awareness of things that can affect anybody's life, to gain knowledge as well as recognize objects in nature $_{(OST)}$[2] (as reported in a survey questionnaire). One of the objectives of teaching Biology at the higher secondary level is also to understand the significance of Biotechnology in real-life context and take cognizance of the strides made in Biology $_{(OST)}$ (as reported in a pre-observation questionnaire). She considers scientific literacy to be an important aim of science education $_{(OST)}$ and defines it as an ability to find out reasons behind day-to-day happenings. An example that she quoted in this regard was of Amniocentesis, wherein the students can be made aware of the 'right' use which is for an early detection of chromosomal disorders and the misuse of the same technique for female foeticide. Also, determining the right marriageable age for girls and preventing the incidence of child marriage and forced marriages. $_{(KCur+KTS)}$

6.3 Understanding About Nature of Science (NOS)

Saira feels that History of Science should form an important component of sciences as it helps in understanding how the scientific knowledge develops and also what could be the future consequences of any development in science and technology $_{(KNOS)}$. According to her, scientific principles are universal in nature and can be applied equally well at any place and at any time.

> Scientific principles are universal in nature, the nature of light does not vary here in India or any place in the world. –Neither do the laws of Mendel change since it is based on actual observation it has a universal appeal $_{(KNOS)}$.

The present teacher holds the opinion that scientific knowledge is reliable in terms of its universal applicability but here she did not mention the limitation of 'observation' as an evidence to prove or disprove a particular scientific phenomenon; since observation can never be completely objective and free from observer's biases, her background, culture, worldview, etc. (Chalmers, 1999). This is how Science is also subjective in nature and scientific theories are subject to modification deeming the earlier theories redundant.

---

[1]Source: Survey Proforma (Appendix C).

[2]The present chapter involves the use of certain codes, e.g. KTS, OST, KNOS, etc. that refer to teacher's knowledge domains (for detailed reference see Appendix F).

6.4 General Notions About Ethics in Science

The present teacher defines ethics in her own terms as "ETHICS provide us with a GUIDELINE to follow what is RIGHT as well as to distinguish between RIGHT and WRONG" $_{\text{(KET)}}$. This was supported by her comment on the statement, "Future citizens will need to be informed consumers of technology, science, sociology and ethics".

> I agree that they should be aware about the consequences of each and every technology and should go for technology that is more environment friendly. $_{\text{(OST)}}$

The inclusion of ethics within science curriculum is well supported by the present teacher especially when she said: "Ethics has to find a place in science $_{\text{(KET)}}$ as at some points the ethical issues have been raised and dealt with but is not linked up explicitly within each chapter" $_{\text{(KCur)}}$. She feels that these issues are of ethical relevance as they are impacting the human race directly and most of the times, in a negative way.[3] $_{\text{(KET)}}$

6.5 Ways of Negotiating Ethical Issues in Biological Sciences

In order to understand the pedagogical approach adopted by the present teacher to deal with ethical issues, some classroom episodes were recorded along with a critical analysis towards the end of each episode so as to bring out the ethical discourse inherent within a particular teacher–student interaction.

The chapter on 'Reproductive Health' is beset with many topics of ethical concerns such as amniocentesis, Medical Termination of Pregnancy (MTP), In Vitro Fertilization (IVF) and other Assisted Reproductive Techniques (ARTs). The following excerpts from the present teacher's classroom on the various topics reveal her own ways of confronting the ethical issues embedded therein.

Box 6.1 Case Study on Amniocentesis

Teacher: *(narrating an instance)* "A doctor by profession gave birth to twin daughters and was very happy. In-laws immensely disappointed on hearing the news and tried every bit to kill those two girls. Once the mother-in-law even pushed the babies onto the stair case, however the girls and mother were rescued by her father." Saira also referred to the recent series hosted by actor Amir Khan called 'Satyamev Jayate' which focused on female foeticide. All such cases clearly point towards only one thing, i.e. immense craving for a male child as a result of which the sex ratio is declining every passing day, and

[3]Source: Questionnaire.

in such disturbed state of affairs such medical technology as 'Amniocentesis' is promoting this female foeticide. The actual benefit of this technology was also being highlighted by the teacher as "Amniocentesis is a technique to identify fetal genetic disorders by taking out few cells from the amniotic fluid surrounding the growing fetus that contains some fetal cells as well" $_{(SMK)}$. Saira mentioned about this malpractice being rampantly followed in certain parts of India by quoting a brief **case study from Haryana** $_{(KTS+KCon)}$.

"In rural sectors such as Haryana and some parts of Punjab, females are being held responsible for giving birth to the girl child and are being punished and mistreated for the same. They have to undergo this procedure of Amniocentesis each time they conceive and in case it is a female child, they have to undergo abortions or forced MTPs (Medical Termination of Pregnancy), sometimes even in the second or third trimesters when it becomes extremely hazardous for the mother's health and life. As a result of this the region has a highly skewed sex ratio and the men in this region often have to marry females from other states."

The narration of the above case (refer to Box 6.1) by the teacher clearly depicts the magnitude and seriousness of the issue. However, some of the implicit ethical questions that remained unaddressed were as follows:

- Why the female is always blamed for giving birth to the girl child? (Is it just due to lack of awareness or a pretext to make her feel small and vulnerable?)
- Do parents have any right to determine the sex of the unborn baby?
- Why in this country are females always the victims of such procedures just because nature has assigned them this task of giving birth?
- Why are males given so much preference and females regarded as a fairer sex?
- Why is this hidden craving and desire for the male child so intense that one is ready to kill the growing female foetus in the womb?

Box 6.2 Classroom Discourse on Birth Control

(Discussion on the rising population and reasons for the same, such as better medical facilities. Quoting situation in China which is following one-child norm and Sweden where there are incentives for those who bear more than one child)

Teacher:	Every sixth person in the world is an Indian. Is it good? What could be its impact? (Consequentialist) $_{(KArg+KCon)}$.
Student 1:	Dearth of resources.
Student 2:	Limited employment opportunities.
Student 3:	Shortage of medical facilities.

Student 4: Rise in the percentage of illiterate population.

Teacher: Due to lack of awareness amongst the uneducated people living in the rural sector as compared to the urban sector where the people are aware about various methods of contraception and understand the importance of small families, this will lead to a disbalance in the population, which would mean uneducated population will be more as compared to the educated lot. This will directly affect the economic development of the nation as a whole and drive our nation towards poverty, unemployment and all those social evils that we vanquished in the recent past. Also, large population size contracts epidemics such as dengue, malaria, swine flu, etc.; therefore, the population needs to be controlled for creating a balanced and well-developed society (Consequentialist) (SMK+KCon+KArg+KTS)

From the above discussion (refer to Box 6.2), the teacher clearly puts forth logical arguments against population explosion and some of its repercussions on the society. This gives rise to another topic of debate, i.e. use of Contraceptives, wherein the present teacher dealt with both, the social and the technical aspects of using contraception as a way of controlling the population. The social aspects include, taboos associated with their use, generating awareness among people of the importance and benefits of small families and family planning as well as consequences of keeping large families. Here an ethical argument raised by the teacher was as follows.

> There are some religions that prevent the use of any contraceptive or family-planning measure, as they feel that children are God's most precious gifts and one has no right to kill a growing fetus which is regarded as a sinful act. (KArg)

This created disequilibrium in the class for some time and an ethical dilemma where personal or societal choices interfere with religion. Whose ethics should be followed and why? However, no further elaboration was done on this topic, and the teacher then switched to some of the side-effects of using various contraceptive methods (refer to Box 6.3).

Box 6.3 Classroom Discourse on Contraceptive Methods

Student 1: Do these contraceptive pills have any side effects?

Teacher: Yes, their repeated usage can cause irregular menstruation, excessive bleeding, mood swings and tender bones. What was the objection raised with the i-pill? (KArg)

Student 2: It is also a kind of feticide.

Teacher: The emergency contraceptive pills such as 'I'-pill is actually abortifacient and prevents conception at the level of fertilization as it interferes with implantation. These kinds of pills also encourage pre-marital sexual relationships. (SMK+KArg+KCon)

Teacher:	Are the Surgical methods of contraception reversible?
Student 3:	No.
Teacher:	They can be reversible if the tubes are ligated back but the operation is not 100% successful

The teacher in the above classroom episodes (refer to Box 6.2 & 6.3) tries to bring to the fore some of the ethical issues associated with the use of contraceptives and managed to highlight their importance in maintaining the population size of a country. But, some other ethical issues associated with the use of contraceptives, such as their excessive marketing leading to promiscuity were kept at bay. keeping in mind the sensitivity attached to the issue (as told in a post-observation interview). There is still some amount of dilemma regarding the surgical methods of contraception as they tend to be irreversible.

Box 6.4 Classroom Discourse on Assisted Reproductive Techniques (ARTs)

Teacher:	Who was the first test-tube baby?
Students:	No responses.
Teacher:	She was a girl named Louise Brown. What is the technique that is employed in the production of test-tube baby?
Student 1:	In vitro fertilization.
Teacher:	The fertilization takes place in the test tube where the ovum from the mother and sperm from the father are fused resulting in a zygote. After the attainment of eight-celled stage, it can then be transferred into the womb of the same female (mother) or another female (surrogate mother) in which case it is called as Zygote Intra-fallopian Transfer (ZIFT), or can also be transferred at the trophoblastic stage. (SMK)

"This In vitro fertilization and Surrogacy has now taken the shape of a booming industry, especially in the developed countries, where people from the affluent families hire women from developing nations who are ready to provide their womb on rent in lieu of huge sum of money and other incentives." (Utilitarian)

[After narrating the instance, the teacher raised a number of questions to the students.]

- Is it fair to reduce human beings to a machinery of producing children? (utilitarian) (KArg)
- What about the emotional state of the (surrogate) mother who gives birth to the child? (KArg)
- Will this technique bring about disturbances in the mother–child bond? (KArg) (Consequentialist)

- Who is to be named the real mother of the child? (KArg)
- What kind of bonding will the child share with the actual mother (who donated the ovum)? (KArg)

However, the other ethical issues attached to the use of contraceptives, such as interference with Nature's rules and some religious arguments, were largely ignored. Although the teacher did raise the issue of contraceptives as abortifacient, the whole debate on 'where does life begin?' was missing from the classroom discussion.

From the above classroom discourse (refer to Box 6.4), it is evident that the present teacher is able to raise some ethical issues related to the technology of In Vitro Fertilization (IVF). However, the students were unable to generate any argument from their side. The reason for this could be lack of understanding about ethical issues or inability to view the issues related to Science and Technology from multiple dimensions. Similarly, in another topic on GMOs (Genetically Modified Organisms), the teacher could initiate a debate in the classroom and tried to generate students' opinions on the issue (refer to Box 6.5).

Box 6.5 Classroom Discourse on Genetically Modified Organisms (GMOs)

Teacher:	What do you understand by 'Ethics'?
Student 1:	Norms of a society.
Teacher:	A set of ideas/beliefs/values that guide a society or functioning of a society. $_{\text{(KET)}}$ So, linking the ethics with the applications of Biotechnology such as production of GM Crops, Transgenic plants and animals, what could be some of the ethical issues that could be raised?
Student 2:	Exploitation of animals on a large scale.
Student 3:	Animals being used in drug experiments and held captive. (Ethic of Care)
Teacher:	Animals are being used as "living laboratories". $_{\text{(KArg+SMK)}}$ What else?
Student 4:	Excessive exploitation of an animal can put a species into the endangered zone.
Teacher:	What about GM crops? GM crops may cause allergy, we actually don't know or not sure about what could be the harmful effects of such crops on humans. (Consequentialist) $_{\text{(KArg)}}$
Student 5:	Cultivation of such crops may affect soil fertility. (Consequentialist)
Student 6:	But effects can only be seen once after consumption.

Teacher:	For regulation, monitoring, assessment as well as impact study of GM crops, a separate body has been created with the name of GEAC (Genetic Engineering Appraisal Committee). However, GM crops can be compared with a "Pandora's Box" which is a closed box and wherein we do not know what the consequences could be (good/bad). Are we trying to play God? What could be some of the after-effects on plants, animals, insects, etc.? (Consequentialist) $_{\text{(KArg)}}$ In a 'Biological Warfare', some organisms are deliberately created by means of genetic modification so as to cause harm. (Principle of non-maleficence) $_{\text{(KArg)}}$
Student 7:	Many a times, during a war period, soldiers of the enemy group are fed with bacteria-injected food.
Teacher:	In 'Biopiracy', using genetic information related to a particular gene for creating a better gene or for integrating it with another sequence of genes (Recombinant DNA technology) takes place but without proper authorization. This usually happens when one country (most of the times an underdeveloped/developing country) is rich in terms of resources and biodiversity therefore accounts to a diverse gene pool and the other country (most of the times a developed nation) has the best technological expertise, so that they can exploit the resources from the other countries. Example can be of 'Basmati Rice'. The U.S has patented it. $_{\text{(SMK+KTS+KArg+KCon)}}$
Student 8:	Did they take permission from the country whose resources they were using?
Teacher:	No, they did not. This resulted in a lot of controversy and protests from the farmers' and agriculturists' side.
Student 9:	There should be a MoU signed by both the countries so that the contribution and efforts of both the countries is being taken care of and acknowledged (Principle of Justice).
Student 10:	The Biopatent should only be for a limited period.

In the above classroom discourse, the various ethical issues such as those related to animal exploitation, use of animals in research purposes, effects of GM crops on human health, soil fertility and environment, issue of biowarfare and biopatents were raised. However, besides listing the ethical issues, the discussion on the same seemed to be limited. Some real cases and examples could have substantiated the arguments and enhanced students' understanding in a better way. Although, the students raised arguments with respect to GMOs and are getting into the complexity of these issues, but their arguments are weak as they are not able to support them with evidences and theoretical backing. It is evident that the interest level of students is also high while discussing about these issues in the classroom. The class is a case in point depicting equal opportunities being distributed to all the students vis-à-vis classroom participation and opinion sharing.

Summary of Classroom Processes

S.No.	Classroom processes	
1.	Introduction to the Topic	Connecting with daily life incidents Plainly introducing the topic (No Background)
2.	Kind of examples/analogies used	From daily life and personal experiences From television and other media From the textbook (if any)
3.	Number of ethical arguments raised	5–8 per class
4.	Type of ethical arguments raised (along with some evidence)	**Utilitarian** ("This In vitro fertilization and Surrogacy has now taken the shape of a booming industry, especially in the developed countries, where people from the affluent families hire women from developing nations who are ready to provide their womb on rent in lieu of huge sum of money and other incentives.") **Consequentialist** ("GM crops can be compared with a 'Pandora's Box' which is a closed box and wherein we do not know what the consequences could be (good/bad). Are we trying to play God? What could be some of the after effects on plants, animals, insects, etc.?")
5.	Treatment given to ethical arguments	Listed and discussed
6.	Students' contribution to ethical arguments	Students participating and contributing in discussion "Do these contraceptive pills have any side effects?" "It is also a kind of feticide."
7.	Classroom time devoted to the teaching of ethical issues	15–20 min per class
8.	Kind of resolution reached	A dilemma created amongst students
9.	Major pedagogical approach adopted	Discussion and debate Quoting vignettes, personal experiences, reports from the newspaper, magazines, etc.
10.	Kind of ethical understanding aimed at by the teachers	Ethical Dilemma level of understanding

6.6 Understanding with Respect to the Ethical Issues in Biological Sciences

In-depth interview sessions with the present teacher helped in generating information, opinions and beliefs about some of the ethical issues related to Biological Sciences. The responses have been lifted directly from the interview transcripts and have been presented here in a tabular form along with the critical analysis (refer to Table 6.1).

From the responses, it can be said that the teacher has some content knowledge with respect to the ethical issues in Biological Sciences but the same is not sufficient to generate an authentic ethical understanding among the learners about issues of ethical relevance related to Biological Sciences. She is able to elaborate upon some of the ethical issues associated with say animal experimentation, amniocentesis, contraception, IVF, HGP and conservation of natural resources, however, in the case of GM crops, GMOs, she seems to have an ambivalent viewpoint and is not sure about the technology and its impact on humans. It was observed that wherever she provides evidence to support her argument, she fails to mention the source of that evidence which is very important for bestowing validity to her arguments. An argument analysis will help in analysing the quality of the teacher's arguments in a better way.

6.7 Argument Analysis

The analysis of the teacher's arguments reveals that her arguments are well-structured, consisting of sound claims, with valid reasons, warrants and backing (refer to Appendix H). In some cases, there are also present strong rebuttals depicting her in-depth knowledge of the topic (refer to Table 6.2).

The argument analysis of the present teacher's arguments reveals that she took recourse to the deontological, consequentialist and utilitarian theory most of the times as a 'backing' and a strong base for her claims. For most of the topics, the teacher was able to provide a strong rebuttal which indicates a 'good quality' argument, as she is aware of the different facets of a particular technology. The data or evidence are also contextualized to the present situation and often link with the larger sociopolitical and economic milieu,

> In Andhra Pradesh the Government fell because of this Bt-cotton and also led to huge number of farmer suicides.
>
> In vitro fertilization and Surrogacy has now taken the shape of a booming industry.

However, she did not mention the source from where these evidences and reasons were quoted and rather appear to be based more on a popular notion or staunch belief.

Table 6.1 Analysis of teacher's arguments based on ethical theories

S. No.	Topic	Ethical issues	Case's SMK/Opinions/Beliefs/Metaphors	Related ethical theory
1.	Animal Experimentation	1.1. Use of animals in research and study 1.2. Pain and suffering to the animal 1.3. Wastage of animals	"Actually at school level also it was a mistake to ban animal experimentation totally, because you know when a child sees as to how a heart actually looks like then it gives a different feeling, I think even the 3-D figures also do not give such a kind of a feeling." (TB)	Utilitarian
			"We should focus on what is the treatment being given to animals after experimentation and there should be some kind of animal hospitals to take care of them."	Ethics of care
			"We have to do experiments on animals for progress." (TB)	Utilitarian
2.	Amniocentesis	2.1. Sex determination 2.2. Female foeticide 2.3. Craving for the male child	"Amniocentesis should not be used for sex determination." (SMK+TB)	Deontological
			"If done for positive reasons, then yes amniocentesis should be done as it helps in identifying many congenital diseases." (SMK)	Utilitarian
			"When the question is of one's own child then people have a craving for a male child because he is the ultimate possessor and heir of the family's name, even if he kicks off his parents in their old age." (TB)	Feminist Ethic
3.	Contraception	3.1. Religious ban on contraceptives 3.2. Contraceptives can be abortifacient 3.3. Fringing right to birth and right to life 3.4. Conceptus as a potential human being	"Yes, some religions like Catholics' and even the Popes are against the use of contraceptives because they think that children are God's gift." (SMK)	Religious ethics
			"If a family is not in a position to support another individual in the family then what will such a potential human being do in poverty where even there are no basic amenities available." (TB)	Consequentialist
4.	In Vitro Fertilization	4.1. Safety of the technique in terms of the mother's health. 4.2. Wastage of embryos 4.3. Estranged relationships between child–parent, child–sibling, child–society, etc.	"It's like a boon to childless couples" (SMK)	Utilitarian
			"Surrogacy or sperm/egg donation may be somewhat controversial but if both the parents are okay and just the fertilization is being done in vitro then it is fine." (SMK) "I think that at the single cell stage it has no life, here comes philosophy, you have to define where life is?" (TB)	Ethics of human right

(continued)

Table 6.1 (continued)

S. No.	Topic	Ethical issues	Case's SMK/Opinions/Beliefs/Metaphors	Related ethical theory
5.	GM Crops	5.1. Do we really need GM? 5.2. After-effects of GM on human health and environment 5.3. Accessibility and means to buy GM crops/seeds 5.4. Is organic food a better option? 5.5. Gene patents and biopiracy	"Genetically modified banana could cure diarrhoea in children and then vitamin A enriched rice, but still I feel that who would go for vitamin A enriched rice when you have carrots which are a rich source for the same. So, may be there was no need for Genetically Modified crops." (SMK+TB)	Utilitarian
			"May be pest resistance could be one of the reasons for the introduction of GM technology." (SMK+TB)	Positive Utilitarian
			"Bt-cotton I don't think it is successful because in Andhra Pradesh the Government fell because of this Bt-cotton and also led to huge number of farmer suicides." (TB) "GM crop is not for a country like India where there is mass scale corruption although testing is being done, but how we don't know." (TB)	Negative Utilitarian
			"I don't think that a common man would even be able to afford GM crops at such exorbitant prices." (TB)	Principle of Equality
			"There are many issues such as what if the bacterial toxin affect the human system and some microbes in our body might take up that toxin to develop antibiotic resistance." (SMK)	Consequentialist
			"I think biopatent is a must because any country needs to preserve its resources and reserves against invasion." (SMK+TB)	Deontological
6.	Human Genomic Project	6.1. Principle of privacy and confidentiality of genetic information 6.2. Playing with nature's diversity and gene pool 6.3. Issue of risking human life 6.4. Issue of informed consent. 6..5. Gene patent	"I think at this level the focus should be on understanding what HGP is all about, advantages of genome project in say future of medicine, and emphasis I think should be more on how this human genomic project can be used to cure diseases, and thus more of advantages and some knowledge about disadvantages." (OST+KTS)	Positive Utilitarian
			"Insulin can also be synthesized; most of the metabolic disorders can be treated by using gene therapy, once you have the sequence of genes that is responsible for producing particular enzymes." (SMK)	Utilitarian
				Consequentialist

(continued)

Table 6.1 (continued)

S. No.	Topic	Ethical issues	Case's SMK/Opinions/Beliefs/Metaphors	Related ethical theory
			"Once the tool is in the hands of miscreants it can result in biological warfare, or certain organisms can be specifically created which can say attack humans leading to mass destruction, people can actually engineer such organisms." (SMK+TB) "There needs to be stricter control and regulation only then can the technique be really useful."	
7.	Stem-cell therapy and Human Cloning	7.1. Issue of potentiality 7.2. Issue of Human Rights	"Stem cells can also be harvested from the	Utilitarian
			placenta, which is a waste otherwise." (SMK) "If the stem cells are derived from the embryos directly then off course people may feel uninhibited going for MTPs and abortions, which could mean posing threat to humanity as a whole." (SMK+TB)	Ethics of Human Rights
8.	Conservation of Natural Resources	8.1. Overconsumption of Natural Resources 8.2. Responsibility towards future generations 8.3. Principle of equality and equal distribution of resources	"It should be used judiciously, tell them the consequences of depletion of resources, for instance, earthquakes are happening due to procedures such as indiscriminate mining, etc. So, make them aware of both the positive and the negative aspects." (SMK+OST+KTS+KCur)	Ethics of Conservation
			"Actually the developed countries are to be blamed as they are the real culprits who have been using technology and its products for so long, and we are right now at a developing stage and therefore need to set up industries." (SMK+TB)	Utilitarian
			"Children at this stage should be taught to conserve natural resources from a broad utilitarian perspective, they should be told to appreciate the natural resources, appreciate a water body or appreciate forests and other things." (SMK+OST+KTS+KCur)	Utilitarian

Source In-depth interviews with the case

Table 6.2 Argument analysis based on Toulmin's method argument analysis (Average score = 3.6)

S. No.	Argument topic	Data	Warrant	Qualifier	Backing	Rebuttal	Claim
1.	Animal Experimentation (Level = 2)	At school level also it was a mistake to ban animal experimentation totally	You know when a child sees as to how a heart actually looks like then it gives a different feeling, I think even the 3-D figures also do not give such a kind of a feeling	There should be some kind of animal hospitals to take care of them	Ethics of Care Utilitarian		We have to do experiments on animals for progress.
2.	Amniocentesis (Level = 3)	People have a craving for a male child because he is the ultimate possessor and heir of the family's name			Deontological	Amniocentesis should be done as it helps in identifying many congenital diseases	Amniocentesis should not be used for sex determination
3.	Contraception (Level = 5)	Every sixth person in the world is an Indian. China follows a one-child norm. Large population size contracts epidemics such as dengue, malaria, etc.	Rising population will directly affect the economic development of the nation as a whole	If a family is not in a position to support another individual in the family	Consequentialist	Repeated usage of contraceptives can cause irregular menstruation, excessive bleeding, mood swings and tender bones. These kinds of pills also encourage pre-marital sexual relationships	Population needs to be controlled for creating a balanced and well-developed society

(continued)

Table 6.2 (continued)

S. No.	Argument topic	Data	Warrant	Qualifier	Backing	Rebuttal	Claim
4.	In vitro Fertilization (Level = 3)	In vitro fertilization and Surrogacy have now taken the shape of a booming industry	At the single cell stage it has no life	If both the parents are okay and just the fertilization is being done in vitro then it is fine	Utilitarian	Is it fair to reduce human beings to a machinery of producing children?	It is like a boon to childless couples
5.	GM Crops (Level = 4)	Genetically modified banana could cure diarrhoea in children and then vitamin A enriched rice In Andhra Pradesh, the Government fell because of this Bt-cotton and also led to a huge number of farmer suicides	Although testing is being done, how we do not know I do not think that a common man would even be able to afford GM crops at such exorbitant prices Bacterial toxin affects the human system and some microbes in our body as it might take up that toxin to develop antibiotic resistance		Consequentialist	Maybe pest resistance; could be one of the reasons for the introduction of GM technology	GM crop is not for a country like India where there is mass scale corruption Bt-cotton I do not think it is successful
6.	HGP (Level = 5)	Insulin can also be synthesized	Most of the metabolic disorders can be treated by using gene therapy once you have the sequence of genes that is responsible for producing particular enzymes		There needs to be stricter control and regulation, only then can the technique be really useful	Once the tool is in the hands of miscreants, it can result in biological warfare, or certain organisms can be specifically created which can say attack humans leading to mass destruction	Emphasis I think should be more on how this human genomic project can be used to cure diseases

6.8 Sensitivity Towards Ethical Issues

During the in-depth interview sessions, the present teacher made use of emotional metaphors[4] a number of times which clearly depict her sensitivity and concern for a particular issue (refer to Table 6.3).

Table 6.3 Saira's emotional metaphors

Planes	Emotional metaphors
Individual	**Animal Experimentation** When a child sees as to how a heart actually looks like then it gives a different '*feeling*', I think even the 3-D figures also do not give such a kind of a feeling. There should be some kind of animal hospitals to take '*care*' of them **Amniocentesis** If done for 'positive' reasons, then yes amniocentesis should be done as it helps in identifying many congenital diseases. They have to undergo this procedure of Amniocentesis each time they conceive and in case it is a female child they have to undergo abortions or '*forced*' MTPs (Medical Termination of Pregnancy), sometimes even in the second or third trimesters when it becomes *extremely hazardous* for the mother's health and life. **IVF** This in vitro fertilization and Surrogacy has now taken the shape of a *booming* industry, especially in the developed countries Even if it is developed from a fertilized ovum then also it is just at a one- to two-cell stage and I do not regard it as a potential human being. So, according to me, it does not matter even if it is *wasted.* **GM crops** GM crops can be compared with a "*Pandora's Box*" which is a closed box, wherein we do not know what the consequences could be (good/bad). **Stem-cell Therapy** If the stem cells are derived from the embryos directly, then of course people may '*feel uninhibited*' going for MTPs and abortions, which could mean posing '*threat*' to humanity as a whole. **Conservation of Biodiversity and Resources** Maybe it should be used '*judiciously*' There has always been a '*conflict*' as developed countries always blame the developing countries for their increasing levels of carbon footprints They should be told to '*appreciate*' the natural resources, appreciate a water body or appreciate forests and other things. Because they are the future country or nation as a whole, so if they are '*sensitized*' then the things will work out better.
Relational	
Sociopolitical	Text always builds a 'positive' picture of science and technology

[4]Emotional Metaphors indicate how someone feels about events, objects or people in his/her life. Metaphorical language is said to be an extraordinarily powerful tool through which teachers express the meaning of their experiences in teaching and much of their pedagogical knowledge which teachers hold (e.g., see Briscoe, 1991; Clandinin, 1986).

6.9 Idea About the Curriculum at Senior Secondary Level (with respect to inclusion of Ethical Issues)

The teacher does not seem to be very satisfied with the present curriculum and its structure, especially with respect to the treatment given to ethical issues. This was evident when she said:

> No, these ethical issues are not given their due importance and do not occupy much of the space in the textbook. And since these are not being addressed in the textbook adequately therefore their treatment at the transaction level is also very fleeting. Had they been taken care in the textbook or mentioned in the curricular guidelines then surely their treatment and importance would have been very different. $_{\text{(KCur)}}$

Thus, the above statement made by the teacher calls for well-organized textbook content where these ethical issues receive their due space. She is of the opinion that guidelines should be provided in the curriculum to deal with these issues in a better manner with the students. She strongly feels that ethics should be made an important and indispensable part of sciences and technology.

> Ethics comes in when we need to take decision regarding the use or misuse of a particular technology. $_{\text{(KET)}}$ For instance, during building of a dam at one place, so many populace is being evacuated from there and not even bothered to give them alternative accommodation, this raises both an ethical as well as human rights issues pertaining to development at the cost of human survival. $_{\text{(SMK+KET+KTS)}}$

Thus, the present teacher prefers an application level of understanding by incorporating ethics in sciences so as to take practical decisions with regard to the use of any technology. However, a stark statement which seemed to be a contradiction to whatever has been proposed till now was when asked about the impact of recent science and technology disasters such as Nuclear reactor peril in Japan, or Endosulfan tragedy in Kerala on classroom teaching, she said:

> No they do not impact. As we need to deal with only those concepts that are given in the text, and the text always builds a positive picture of science and technology. $_{\text{(OST+KCur)}}$

The above statement in a way gives a peep into the psyche of the teacher who limits her teaching to the textbook content and topics only and somewhere alienates it from the daily life context. The aspect of ethical issues which has been touched upon in the present curriculum was explained by the teacher as,

> Mostly the technical aspects are being dealt with but do not touch upon some of the emotional or ethical issues, which could have been done by addressing some real cases in the textbook. $_{\text{(KCur)}}$

The limited understanding as well as focus on these issues as explained by this teacher is attributed to their weightage in the exams that focus more on concepts and facts. This is the reason why no specific assessment procedure has been designed for these topics and areas.

6.10 Lesson Planning

The responses gathered from the planning sheet can be presented as follows.

Topic 1 (any ethical issue): Human Genomic Project (HGP) **Class Time**: 30 min			
Planning (Activity/ Experiment/ Demonstration/ Discussion/etc.)	**Mode of Execution** (Process involved along with time required)	**Desired Outcome** (Intended Objectives)	**Assessment**
Reading from reference books, internet and Google search, YouTube, newspaper articles, etc.	Discussion with the students	1. To create awareness amongst students. (OST) 2. To develop scientific literacy. (OST) 3. To generate awareness about ethical issues in Science. (OST)	1. Students' Feedback and responses (KA)
Topic 2 (any topic other than ethical issues): Principles of Inheritance **Class Time:**			
Planning (Activity/ Experiment/ Demonstration/ Discussion/etc.)	**Mode of Execution** (Process involved along with the time required)	**Desired Outcome** (Intended Objectives)	**Assessment**
Reading from the reference books. Demonstration of a monohybrid and a dihybrid cross	Demonstration of crosses using Rajma and Lobhia seeds. Taking examples from real-life situations, example, eye colour, blood groups, etc.	Visualization of the process of inheritance	Making pedigree chart/tree depicting the patterns of inheritance of blood groups or any other genetic trait

Thus, from the planning chart, it is evident that no specific planning is done to deal with the ethical issues (HGP) as compared to the planning done for the teaching of other topics. There are no specific activities being planned or designed in order to deal with the ethical issues as has been done for the other topic. The only pedagogical strategy suggested by the teacher for fulfilling the objectives of the lesson is discussion. The course of discussion, the kind of questions to be asked, the role of teacher, role of students, rules for participating in the discussion, methods of assessment, etc. were not factored in the lesson plan. This indicates lapses in teacher's planning, effort and understanding to choose the right kind of strategy for effectively addressing these issues in the classroom. This can also be taken as a lacuna on the part of teacher preparation programmes that do not focus on these issues of ethical relevance either on subject matter knowledge or on developing interlinkages between subject matter and pedagogy.

As per the planning for the class, the present teacher said in a pre-observation interview that not enough reading material is available with respect to these newer topics in the text such as Bacterial and Yeast Artificial Chromosome and technical details about human genome $_{(KCur)}$ which makes her feel incompetent to deal with students' doubts and difficulties in the classroom. This was very much evident in the classroom observation as at some points, the present teacher was unable to cater to students' difficulties when some questions raised by the students were truly challenging such as, "Why is HGP (Human Genomic Project) such an expensive project?", "Are the benefits of HGP equally shared amongst all?", and "Why HGP if there are so many ethical and legal issues attached to it?". The answers to all such questions seemed to be difficult for her and after the class in a post-observation interview she admitted that she was not well-prepared for all these students' queries and doubts. The present teacher also highlighted that no practical experience at this stage also makes the understanding of these topics abstract for the students.

> As there are no practicals at this stage for them related to HGP, therefore merely conceptual details will create more abstraction in their minds and they will feel find it difficult to connect technology with their real-life context. (Post-observation interview) $_{(KSU+KCon+KCur)}$

6.11 Discussion

The major portion of the present teacher's PCK comprises the SMK, TB and KArg followed by KTS and KCur (refer to Table 6.4 and Fig. 6.1). She believes that ethics should be placed within the science curriculum in the form of ethical issues so as to make the learners not only understand the pros and cons of various technologies but also develop the capacity for ethical reasoning and decision making. While in the classroom, teacher prefers to deal with the issues of ethical relevance by engaging in a discussion and narrating case studies that help in linking the text with the learners' context, wherein students are given complete freedom to participate in the discussion and express their opinion. The diversity of students' arguments on an ethical issue sans teacher's influence is indicative of the fact that there are no power equations working in the classroom, neither any authority of the teacher, hence rejecting the 'banking concept' of education (Freire, 1972). Most of her arguments with respect to various technologies and their use are based on the Utilitarian ethics and the principle of beneficence, which means that she stands in favour of those technologies that are beneficial for the human race and do not pose any harm. The present teacher is aware of both the positive and negative consequences of various Biotechnologies and is able to view their impact from a multidimensional perspective which is evident in her arguments,

Table 6.4 Teacher's knowledge components constituting PCK

S.No.	Teacher's knowledge component (Code)	Evidence (in Classroom Interaction, Discussion, Interview, Questionnaires, Lesson plans, Reflective Fastwrites)	Frequency
1.	Subject matter Knowledge (SMK)	"Amniocentesis is a technique to identify fetal genetic disorders by taking out few cells from the amniotic fluid surrounding the growing fetus that contains some fetal cells as well".	32
2.	Teacher's Beliefs (TB)	"We have to do experiments on animals for progress". "I think that at the single-cell stage it has no life, here comes philosophy, you have to define where life is?"	17
3.	Knowledge of Transactional Strategies (KTS)	"Children at this stage should be taught to conserve natural resources from a broad utilitarian perspective, they should be told to appreciate the natural resources, appreciate a water body or appreciate forests and other things".	15
4.	Knowledge of the Curriculum (KCur)	"At some points the ethical issues have been raised and dealt with but is not linked up explicitly within each chapter".	9
5.	Knowledge of Students' Understanding (KSU)	"As there are no practicals at this stage for them related to HGP, therefore merely conceptual details will create more abstraction in their minds and they will feel it difficult to connect technology with their real life context".	3
6.	Knowledge of Context (KCon)	The teacher mentioned about this malpractice being rampantly followed in certain parts of India by quoting a brief case study of Haryana.	7
7.	Knowledge of Assessment (KA)	Students' feedback and responses	1
8.	Knowledge of ethical theories/ Role of ethics (in general and in Science) (KET)	"ETHICS provide us with a GUIDELINE to follow what is RIGHT as well as to distinguish between RIGHT and WRONG".	6

(continued)

Table 6.4 (continued)

S.No.	Teacher's knowledge component (Code)	Evidence (in Classroom Interaction, Discussion, Interview, Questionnaires, Lesson plans, Reflective Fastwrites)	Frequency
9.	Orientation towards Science Teaching (OST)	"Teaching science at senior secondary stage is to create awareness about things that can affect anybody's life, to gain knowledge as well as recognize objects in nature".	11
10.	Knowledge of Ethical Argumentation (raising arguments/ initiating argumentation/resolving) (KArg)	"There are some religions that prevent the use of any contraceptive or family planning measure, as they feel that children are God's most precious gifts and one has no right to take its life which is regarded as a sinful act".	19
11.	Knowledge about Nature of Science (KNOS)	"Scientific principles are universal in nature, the nature of light does not vary here in India or any place in the world".	15
		PCK_{TOTAL}	**135**

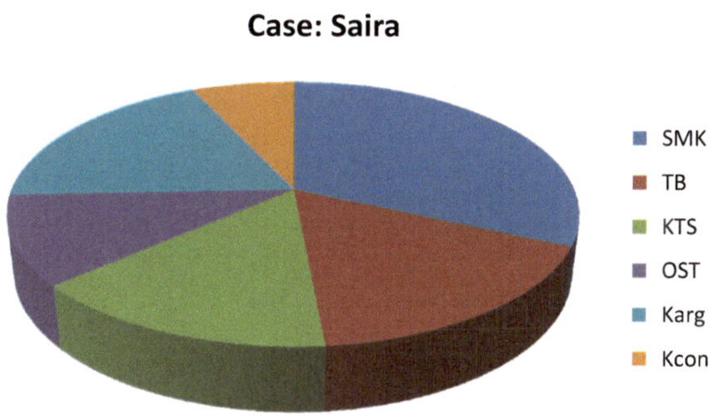

Fig. 6.1 Graphical representation of components of PCK

> Bt-cotton I do not think it is successful because in Andhra Pradesh the Government fell because of this Bt-cotton and also led to huge number of farmer suicides.
>
> I do not think that a common man would even be able to afford GM crops at such exorbitant prices.
>
> There are many issues such as what if the bacterial toxin effect the human system and some microbes in our body might take up that toxin to develop antibiotic resistance.

Thus, the teacher does not hold a positive notion about the introduction of GM crops in India. However, this is not the same for other technologies such as IVF and Stem-cell therapy. Thus, a teacher's idea about 'technology' is based on a Utilitarian and Consequentialist ethic.

6.12 Conclusion

The teaching–learning of ethical issues in Biological Science requires a shift in the perspective from treating Science as objective and value-free to understanding Science as a subjective enterprise beset with values of its own kind. When it comes to technology (which is the application of a scientific principle or law), then a decision has to be made whether a particular technology should be brought into the public domain or not, what could be the possible consequences of such a decision on the society, humans, other living beings and environment. Here, only the scientific way of thinking may not lead to the 'right' decision-making as the technology under question directly impacts the subjective human element. Therefore, it becomes imperative to understand the needs and working of the present society, prevailing norms and behaviours, human values and ethics. An ethical mode of inquiry is pertinent before taking a decision vis-à-vis any given technology so that it does not lead to anymore disasters and devastation for the people and the environment. In the present case, the teacher is aware of the ethical issues associated with the given Biotechnologies and is motivated to transact them in the classroom so as to make her learners aware. From analysing the classroom discourses, it is evident that the present teacher is able to generate an ethical inquiry around these issues. Based on the implicit ethical frameworks, the present teacher guides her students towards ethical decision-making with respect to various Biotechnologies. Although famous psychologists like Jean Piaget and Lawrence Kohlberg have advocated and propounded a stage-wise development of moral reasoning (Piaget, 1932; Kohlberg, 1958), it has been realized by more recent researchers that it is very difficult to allocate a particular stage of ethico-moral development to a particular age-group level, as children have complex lives, diverse experiences, traditions, religion, etc. which lead to differences in their behaviour. A range of parameters that suggest a progression in ethical thinking and reasoning is being proposed that entails the maturity in the thinking process from being a novice towards an advanced level ethical reasoning (Reiss, 2010). This involves the transformation of thinking from being egocentric to following social norms and rules and finally towards applying a more reasoned moral or ethical principle in the given situation (refer to Fig. 6.2). A nuanced ethical understanding is required in situations where there is a conflict between competing ethical theories and principles so as to reach a decision. The change in the thinking pattern also includes looking at a particular issue from a long-term and futuristic perspective rather than a short-term perspective, and researching newer knowledge to combat a problem or difficulty.

Thus, the above figure (refer to Fig. 6.2) clearly depicts that students in the same class may be at varying levels of their ethical reasoning so it is unlikely that all the students will reach similar ethical conclusions or decision-making with respect to ethical issues related to Biotechnology. Therefore, a teacher should listen to different student opinions and should promote a healthy discussion in the class rather than promoting one particular viewpoint.

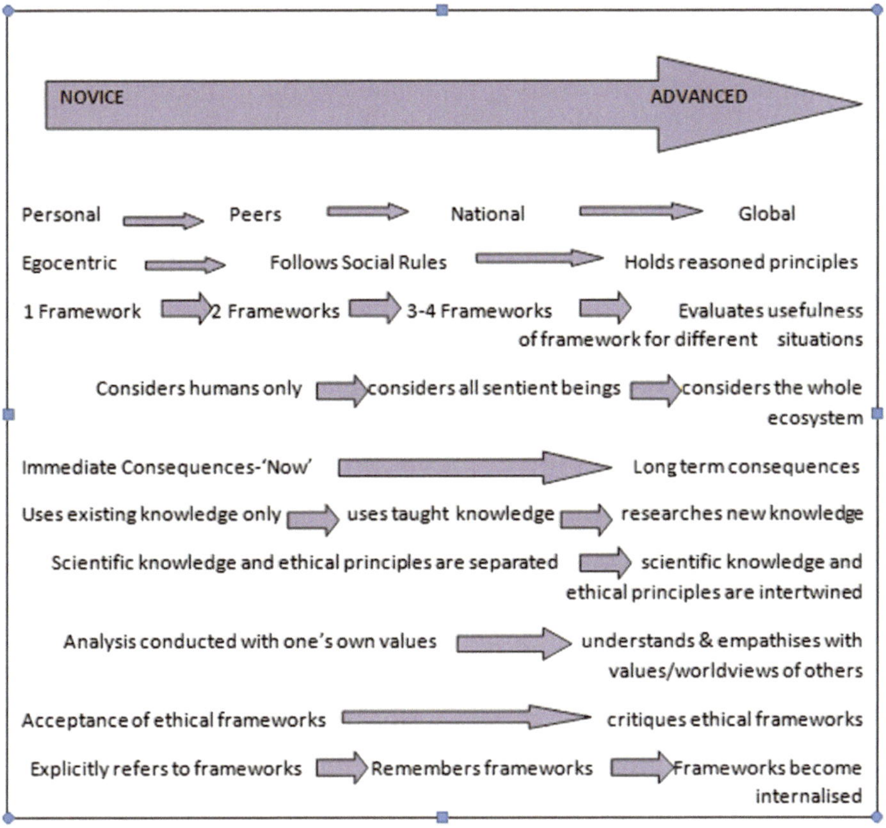

Fig. 6.2 Indicators of progression in ethical reasoning (Reiss, 2010)

References

Briscoe, C. (1991). The dynamic interactions among beliefs, role metaphors, and teaching practices: A case study of teacher change. *Science Education, 75*(2):185–199.

Chalmers, A. F. (1999). *What is this thing called science?* (3rd ed., pp. 19–26). Buckingham: Open University Press.

Clandinin, D. J. (1986). Classroom practice: Teachers' images in action. London: Falmer Press.

Freire, P. (1972). *Pedagogy of the oppressed.* U.S.A: Penguin Books.

Kohlberg, L. (1958). The development of modes of thinking and choices in years 10 to 16. Ph.D. Dissertation, University of Chicago.

Piaget, J. (1932). *The moral judgment of the child.* London: Kegan Paul, Trench, Trubner and Co.

Reiss, M. (2010). Ethical thinking. In A. Jones, A. McKim, & M. Reiss (Eds.), *Ethics in the science and technology classroom* (pp. 1–6). Netherlands: Sense Publishers.

Chapter 7
Promoting Interdisciplinary Thinking for Ethical Understanding—Case Narrative of an Undergraduate Teacher

Abstract The present chapter presents a case study of an Undergrad Biology teacher named 'Meena' and her different knowledge domains related to the teaching–learning of ethical issues in Biological Sciences. These include her idea about teaching, learning, pedagogy, curriculum, students' misconceptions, assessment, views on Nature of Science (NOS), etc., in the context of ethical issues in Biological Sciences. An attempt has been made to document the classroom discourses of the selected classroom situations wherein an interpretive approach to classroom discourse analysis has been adopted. Negotiating ethical issues pertaining to Biological Sciences in the classroom often requires taking recourse to a multidimensional and interdisciplinary way of teaching–learning. This also requires a multidisciplinary understanding of issues at the interface of Science, Technology and Society (STS). Teaching ethical issues pre-empt taking recourse to only one disciplinary approach which would be inappropriate rather limiting to deal with the scope of these issues. The different disciplines provide multiple perspectives on a given Bioethical issue, such as from a scientific, sociopolitical, historical, ethical, legal or economic viewpoint. The present case study is a case in point where the teacher promotes such an interdisciplinary understanding of ethical issues among her learners.

Keywords High school · Biology teacher · Ethical issues in biological sciences · Multidimensional thinking · Interdisciplinary science teaching · Ethical understanding · Pedagogy · Assessment

7.1 Introduction

Meena is an Assistant Professor at one of the reputed colleges in Delhi, India. She holds a Ph.D. degree in Zoology from Osmania University, Hyderabad and a teaching experience of around 10 years at undergraduate level. Her areas of interest in teaching include Physiology, Biochemistry, Immunology, Biotechnology, Biodiversity and Bioinformatics. She uses all kinds of references in her teaching including textbooks, private publication books, e-books, journals, newspapers, web

A. Saxena, *Ethics in Science*, https://doi.org/10.1007/978-981-32-9009-9_7

search, etc., along with self-prepared teaching notes. The teaching–learning approach as advocated by her at the undergraduate level should be a combination of lecture, discussion, interaction and use of audiovisual materials $_{\text{(KTS)}}$.[1] The practical approach to the subject is what she thinks is required at this level, with exploratory activities that provide ample scope for innovation and creativity[2] $_{\text{(OST)}}$.

7.2 Idea About Teaching Biology at Undergraduate Level

Her idea of teaching Biology is to enhance an understanding of the various aspects of relationship between humans and other organisms with their natural environment and application of biological principles to develop various technologies $_{\text{(OST)}}$. An example of this could be bioinformatics also referred to as Computational Biology where Biology is integrated with Computers. Biology as a subject, according to her equips the learners in recording observations, classifying different living organisms by identifying their characteristic features, developing an ecological thinking, problem-solving, reasoning, project management, etc. $_{\text{(OST)}}$. Students learn to answer why? How? When? What? Biology has lots of practical lab sessions which provide uniqueness to the subject[3] $_{\text{(OST)}}$.

7.3 Understanding of Nature of Science (NOS)

The present teacher holds a rigid and erroneous notion about science and its processes, especially when she mentioned scientific knowledge as reliable and infallible and scientific method as a strict procedure of framing a problem statement, devising hypothesis, collecting evidences, experimentation and reaching inference. She feels that results gathered from a scientific experimentation are based on rationality and objectivity of the researcher. However, she seems to have a multidimensional view of technology and science and features that make it complete $_{\text{(KNOS)}}$. According to her, science is art of life and can be very much seen in the day-to-day life. She expressed the ability of sciences in distinguishing between different kinds of realities $_{\text{(KNOS)}}$, and developing a balanced perspective. According to her, while dealing with a controversial topic, a teacher should allow active participation in the classroom and give freedom to her students, so as to express themselves.

> Teacher should be open and allow presentation and questions, noting down, guide students and telling them think out of box, a middle view all the aspects should be shown to

---

[1]The present chapter involves the use of certain codes, e.g. KTS, OST, KNOS, etc. that refer to teacher's knowledge domains (for detailed reference see Appendix A).

[2]Source: Survey Proforma.

[3]Source: Questionnaire.

students, but if going in wrong direction or weird, appreciate the differing opinions, to analyze merits and demerits. Students will start appreciating. (KTS)

7.4 General Notion About 'Ethics' in Science

The present teacher strongly feels that the progress in science directly depends on the 'ethics' that we follow (KET), especially when she said,

> Ethics is not only about right and wrong but also emphasizes on objectivity, integrity, confidentiality, social responsibility, animal care etc (KET). Ethics means morals or the moral correctness of specified conduct (KET). Ethics are not just the table manners rather it defines the social conditions necessary for human beings to thrive.[4] (KET)

Therefore her concept of 'ethics' seems to be deeply embedded within the society and laid norms of behaviour; however, the interlinkages between science and ethics still do not emerge from here.

7.5 Ways of Negotiating Ethical Issues in Biological Sciences

The present teacher usually takes recourse to classroom discussion and argumentation when it comes to dealing with ethical issues in Biology. Some excerpts from the classroom discussion highlighting an ethical discourse between the teacher and her students can be presented as follows (refer to Box 7.1).

Box 7.1 Classroom Discourse on Genetically Modified Organisms (GMOs)

Teacher:	Do you think that there should be some safety measures or something on which we need to be very careful? (Ethic of safety) (KArg)
Students:	No response
Teacher:	[some hint given by Meena- "do you remember when we talked about plasmids we discussed something called as degradative plasmid isolated from pseudomonas, what was so special about them?] (SMK+KCon)

[4]Source: Questionnaire.

Student 1:	Those are the plasmids which have the ability to decompose complex molecules such as oil, etc.
Teacher:	But, then where is the danger? *Although these plasmids are very useful in the treatment and decomposition of oils and can be very helpful in case of oil spills, but at the same time we have to be very careful regarding the biosafety mechanism. (Principle of Non-maleficence)* $_{\text{(SMK+KArg)}}$ Now, what care should be taken?
Student 2, 3, 4:	That these plasmids should not pose any harm to human beings or to the environment.
Teacher:	If by any chance these plasmids reach the oil wells, then what would happen? (Consequentialist) $_{\text{(KArg)}}$
Student 5:	They will clean all the oil wells.
Teacher:	Till now we have been studying about some of the benefits of these Biotechnological techniques, such as Bacterial transformation. But, do you think that there could be certain drawbacks or some safety concerns attached to their use? Have you heard about GMOs? (Consequentialist) $_{\text{(KArg)}}$
Student 6:	These are the genetically modified organisms or crops.
Teacher:	So, crops that are edible, they are genetically modified, which means that a foreign gene has been added to them, do you think that this foreign gene when ingested by us could cause any side-effect on a person's health? (consequentialist) $_{\text{(SMK+KArg)}}$
Student 7:	It can lead to a mutation.
Teacher:	Till now we ourselves are not clear that which kind of DNA material will produce what kind of effect on different individuals. (Consequentialist) $_{\text{(SMK+KArg)}}$
Student 4:	It can also interrupt the food chain.
Teacher:	How can we be so sure that if a transgene is entering our body, it will not cause any altered reaction? (consequentialist) $_{\text{(KArg)}}$ It is quite possible as it is something new which has not been created by nature. Rather, it has been created by human being.
Student 6:	Each one of us is having a different set of genetic repertoire and thus, the effect of a particular transgene can be varied across different sets of population. One may have a positive effect and the other may have a negative.
Teacher:	Yes… any other argument?

Student 7:	If while making a transgenic bacteria containing marker genes which are usually antibiotic resistance genes, then this could lead to the production and selection of some antibiotic-resistant bacterial strains or super-bugs. (Consequentialist)
Teacher:	Yes, and that will mean they will have resistance toward a broad spectrum of antibiotics, on which even a very low to a very high dosages of antibiotics will have no effect and the day will not be far when repeated dosages of these antibiotics will confer more and more resistance to these bacteria as in them the plasmid DNA will interact with nuclear DNA for developing more mutations and resistance. (Consequentialist) $_{(SMK+KArg)}$ That means we should be very sure while making different strains of bacteria, so that they are not left freely in the environment. Otherwise, instead of providing some help, these bacteria could be a cause of a major threat. The most devastating of which would be biological warfare. Any other caution that should be taken care of?
Student 8:	In the case of edible vaccines, there is a difference between the plant system and an animal system, and no one can predict what kind of consequences can be there in the long run. (Consequentialist)
Teacher:	The two systems, i.e., the plant system and animal system, have different set of enzymes. Although, it is a very good idea to begin with the edible vaccines, as one is getting here a dual benefit of filling up one's appetite as well as getting immune to a set of diseases. $_{(SMK)}$ But, we never know how this edible vaccine is going to be metabolised in our systems or what kind of immune response will be generated thereafter. You all know that cancers have become so common these days, and we all have these cancer causing oncogenes present in us. Till this time we are lucky enough as these oncogenes are in the inactive stage, but we never know that which chemical, which transgene could switch them on. (Consequentialist) $_{(SMK+KArg)}$ Therefore, Biotechnology as a subject, it is vast with varied applications, but there are always two sides of a coin, and we should not overlook the other side of the coin, i.e., Biosafety and ethics of this subject. $_{(KET)}$

The above classroom discussion can serve as an exemplar for initiating a healthy exchange of arguments between the teacher and her students. The linking threads for a fruitful classroom discussion as is evident from the present teacher's classroom discourse mainly comprise of the quality of questions she posed to her students. Most of the questions were open-ended and invited varied students' responses, such as the following:

> Do you think that there should be some safety measures or something on which we need to be very careful? (Ethic of safety)
>
> But, then where is the danger?
>
> If by any chance these plasmids reach the oil wells, then what would happen? (Consequentialist)

Although short of explanation, students' responses to these questions were quite imaginative and tested their critical thinking skills. These questions also helped in triggering a thought process in students' mind to grasp and understand the multi-dimensional view of the technology as was evident in their responses. The teacher managed to give more informed and elaborative explanations that helped in clarifying doubts and eliminating many myths and misconceptions in students' minds. Another attribute of teacher's personality that could be witnessed was giving equal respect to all the learners and sharing their responses without criticizing them for giving wrong answers. She tried to portray a balanced picture of technology such as GMO which can be both useful and destructive but needs to be regulated.

The pro and anti-GM arguments created an ethical dilemma among students with regard to the acceptance or rejection of this technology which was also one of the teaching learning objectives so as to develop critical thinking and ethical reasoning for such technologies. Students were also able to raise some ethical arguments which clearly depicted their ethical understanding of such issues. However, no consensus could be reached and no means of resolving a particular problem could be devised.

In another classroom episode on the technique of 'Southern Blotting', the ethical issues discussed were quite general and can be applied to the different techniques in Biotechnology. The main ethical issues discussed by the teacher were as follows:

- Fiddling with the naturally occurring genetic material.
- Use of GMOs as 'Bioweapons'.
- Presence of a foreign gene in the food chain leading to 'biomagnification' and 'bioaccumulation'.
- Genetic engineering and selection of favourable traits and formation of 'genetically engineered babies'.
- Loss of species integrity, etc.

The moot point being emphasized here was that any misinformed emotion and negative human values, such as greed, contempt and egoism, could result in harmful scientific research. Both the teacher and her students were in favour of setting up some rules and norms for preventing any misuse of the technology.

The present teacher is somewhat aware about students' difficulty with respect to comprehending the ethical issues embedded in various Biotechnological applications, and thinks that students face difficulty in understanding the theoretical concepts in dearth of practical exposure. However, narrating a case study with respect to any Biotechnological application, she feels provides a context for understanding the inherent ethical issues. Some such case studies put forth by the teacher during an in-depth interview session have been provided in the boxes placed under (refer to Vignettes 1, 2, 3 and 4).

Vignette 1: A Real-Life Ethical Dilemma

"One of my close friends who was expecting had been diagnosed with a hydrocephalous baby. This happens due to the malfunctioning of the neural tube that fails to close resulting in the entry of the water into the brain hence resulting in an enlarged brain size. Such a baby will be mentally retarded and will also not live for a long time. It is in the law that if such a fetus is detected, then the mother can go for abortion, but in my friend's case, it so happened that the hydrocephalic fetus was a baby boy, and her first child was a girl. Her mother-in-law told that her that 'what are you doing, are you going to kill the baby? Whatever, it may be you have to give birth to this child.' Now, this friend of mine was in a dilemma and called me, she told me that she's committing a great mistake by signing on the abortion papers, then I suggested her that she should not go by hearsay, and think about that fetus, also she should follow doctor's advice. Because giving birth to such a child will not only be risky for the fetus, but for the mother, sibling and for the society at large. Finally She aborted the child, and then after few years, she was blessed with a second girl child and now she's happy. This is a kind of real life ethical dilemma that a person faces."

In the above case, the teacher surfaces the issue of sex-selection and preference for a male child over a female child. It also reveals the hidden patriarchy and biased attitude of the people who prefer a male heir. Here in this case, the male foetus is hydrocephalic and thus is a cause of concern for the mother and the family. The ethical dilemma in the present situation is, whether the mother should give birth to such a child? What if the child develops certain abnormalities and is not able to sustain himself? How would the family react to aborting the male child? Thus, the technology of Amniocentesis offers help in identifying many foetal disorders but could be used otherwise as in the present case to determine the sex of the unborn child which is highly unethical and could lead to disastrous consequences.

Vignette 2: Emergency Contraceptives and Their Aftermaths
Some of the students who are in their teenage, just for the sake of say a party, or a ceremony, or before going on trip, etc., they will just go by hearsay and take any medicine that can delay menstruation without consulting a physician. Now in such a case if they keep on using such medicines repeatedly, then in later life they may have to face many problems related to reproductive health. And I do tell and share all this with my students in the endocrinology paper, and I tell them that contraceptive is nothing but a kind of hormone that is being secreted by our own body. And even if it has to be used then one should always consult a gynaecologist and abide by her prescription and advice. Just for the sake of two day or a three day trip, these medicines can play havoc with the system and result in a disaster.

In the above instance, the teacher tries to portray some of the misuses of contraceptives taken just for the sake of one's convenience. By taking these contraceptives without prior advice can lead to certain health-related problems which often the people are unaware of. Thus, the above experience shared by the teacher helps in making the students cautious about some of the harmful effects of such contraceptives on human hormonal system.

Vignette 3: Supernumerary Embryos in IVF and Emotional Bearings of Surrogate Mother
I know some families that have gone through this procedure of IVF, and every time they were told that there would be chances of triplets or quadruplets, so at one point they were happy that they will have their own child but on the other hand, they were also worried about the fact that what will happen if it resulted in three or four embryos together. And moreover, every other day they were supposed to go through ultrasounds and other tests that were also very painful plus economically draining.

Surrogacy is another issue, I remember an incident where a foreign couple asked a woman in Rajasthan to bear their child and serve as a surrogate mother, so for the sake of huge sum of money, the woman agreed, but later on during those nine months of gestation she developed emotional attachment with the growing fetus, as a result of which she refused to part with the baby even in lieu of money. Therefore, resolving such issues becomes very difficult.

In the above case, the teacher shared some of the procedural complications and after-effects of IVF and Surrogacy. Here the ethical concerns raised by the teacher include the amount of pain and money involved in the procedure. Is it really worth to bear so much just for the sake of having one's own child? What if it results in more than one foetus? Is the family ready to bear the responsibility of more than one child?

In case of surrogacy, a poor woman just for the sake of money agrees to be a surrogate mother to a child to be borne by a foreign couple. Here there are many ethical concerns involved such as, is it ethical to separate the child from her surrogate mother? Is there really any need to go for such complicated procedures when there is an option to adopt a child instead? Raising such issues in the class will definitely sensitize the students towards the use and misuse of a technology.

Vignette 4: A Case of Indian Farmers' Suicide

There has been a rise in suicide cases of the Indian farmers who are most of the times the targets. There could be many reasons as such. They were lured by some big companies that if they make use of these seeds, they will be getting a very good harvest, but that did not happen. Not only that, but even when they wanted to get rid of those crops they saw that the fertility of soil was highly affected, so, it is not just one crop that they have lost but also the fertility of the soil. That soil was the only thing they had for their survival (sympathetically)... and they could not pay back the loans, and had frustration and the Government agencies were not there for their help. No information was given on how the conditions can be improved and recouped, they felt cheated by big companies and also by the Government.

The above case points out the experience of Indian farmers on growing GM crops in order to increase their crop yield, but actually resulted in adverse circumstances. The GM seeds could not thrive on the Indian cropland; as a result, the farmers incurred huge losses and were under debt. This whole issue raises ethical concerns such as why these foreign multinational companies allowed to sell their GM seeds to Indian farmers without even testing their impact on the Indian soil? Is there really a need to plant GM seeds when we are still not sure about their future impacts? Such issues raised by the teacher help in developing critical thinking among students vis-à-vis any technology and acquire the ability of decision-making.

Thus, the teacher has a repertoire of transactional strategies in the form of case studies mentioned above, but all could not be actually transacted and put across to the students in the classroom due to paucity of time. No assessment strategies have been planned by her in order to assess students' ethical understanding with respect to these issues, which gives a clear indication that these issues are not considered as important and are not given appropriate place in the curriculum as other topics.

Summary of Classroom Processes

S.No.	Classroom processes	
1.	Introduction to the topic	Plainly introducing the topic (no background)
2.	Kind of examples/analogies used	From scientific researches quoted in newspapers/ journals From television and other media
3.	Number of ethical arguments raised	>10
4.	Type of ethical arguments raised (along with some evidences)	**Consequentialist** "How can we be so sure that if a transgene is entering our body, it will not cause any altered reaction?" "Genetically modified, which means that a foreign gene has been added to them, do you think that this foreign gene when ingested by us could cause any side-effect on a person's health?" "Any organism or newly formed cell that has the capacity to destroy the environment, play with the sustainability of the environment, or in ecological terms, what could happen? Have you heard about biomagnifications or bio-accumulation?" **Utilitarian** "So such patients can be picked up from the sample population and requisite line of treatment will be suggested to them." **Deontological** "Should there be some rules and regulations or we can play or do anything we wish to do with the natural genetic material provided to us?"
5.	Treatment given to ethical arguments	Listed and discussed
6.	Students' contribution to ethical arguments	Students participating and contributing in discussion
7.	Classroom time devoted to the teaching of ethical issues	>30 min
8.	Kind of resolution reached	Students left with a question
9.	Major pedagogical approach adopted	Quoting vignettes or cases interspersed with discussion
10.	Kind of ethical understanding aimed at by the teachers	Ethical dilemma level of understanding

7.6 Understanding with Respect to the Ethical Issues in Biological Sciences

The present teacher presented a sound subject matter knowledge about the various ethical issues related to some areas of Biology particularly pertaining to Biotechnology. A glimpse of her ethical understanding with respect to ethical issues related to Biotechnology has been presented below (refer to Table 7.1).

It was found that teacher's arguments are mostly based on Consequentialist and deontological ethic which is in favour of human well-being. She is able to raise many ethical arguments with respect to the recent biotechnology and its impact on the society at large. However, her arguments are mostly based on her beliefs about the technology and less often supported by valid scientific facts and reasoning. The sources of the data still did not appear in her responses.

7.7 Argument Analysis

The argument analysis of the present teacher's arguments clearly depicts that she is able to generate logical arguments with respect to ethical issues in Biotechnology (refer to Table 7.2). However, at some points, the arguments lack either a warrant, or a qualifier, or rebuttal, which indicates an incomplete understanding of a particular issue (refer to Appendix H).

7.8 Sensitivity Towards Ethical Issues

The present teacher makes use of some emotional metaphors[5] while dealing with the various bioethical issues, which was evident from some of the statements that she made during the classroom discourse and in-depth interview sessions (refer to Table 7.3). These statements reflected her sensitivity and concern towards various ethical issues.

[5]Emotional Metaphors indicate how someone feels about events, objects or people in his/her life. Researches have also shown that how one's professional knowledge in teaching may be identified and analyzed through the use of one's metaphorical language. Metaphorical language is said to be an extraordinarily powerful tool through which teachers express the meaning of their experiences in teaching and much of their pedagogical knowledge which teachers hold (e.g., see Briscoe, 1991; Clandinin, 1986). Lakoff & Johnson (1980) suggest that in order to understand one's thinking, and beliefs we should study the metaphors they use.

Table 7.1 Analysis of teacher's arguments based on ethical theories

S.No.	Topic	Ethical issues	Case's SMK/opinions/beliefs/metaphors	Related ethical theory
1.	Animal experimentation	1.1. Use of animals in research and study 1.2. Pain and suffering to the animal 1.3. Wastage of animals	"Animal experimentation is very important; it is like soul for the subject such as Zoology. Students should get the hands-on experience; definitely killing the animals for fun is not the case here, so I feel that it is important." $_{(Karg)}$ "Animal experimentation does have an impact in granting skill of surgery which is in the right of human benefit, which I think is more important." $_{(Karg+Kcon)}$	Utilitarian
			"Ethics is equally important, as just because it is an animal we cannot take it for granted, we have to follow certain guidelines but at the same time removing all the dissections and animal experiments from the course structure is a bit ridiculous." $_{(Karg+Kcur)}$	Animal rights
			"I definitely feel that the syllabus can be redefined and use of animals can be reduced wherever it can be, for instance for extracting one particular tissue we are dissecting the whole animal, here pooling can be done, for instance if dissection for viewing the animal system is being done then at the same time if the other teacher is present, the tissues of the same animal can be used for making histology slides." $_{(Kcur+KTS)}$	Principle of 3Rs
2.	Amniocentesis	2.1. Sex-determination 2.2. Female foeticide 2.3. Craving for the male child	"Patriarchy can be a cause. But, I do agree that amniocentesis is equally important if the mother's age is above 35 or so. Because giving birth to an abnormal child will not only be risky for the fetus, but for the mother, sibling and for the society at large. Ethics also says that." $_{(Kcon+SMK+Karg+KET)}$	Consequentialist
			"What I feel is that strict enforcement of the law should be there. People involved in such a cruel act should be punished.… Rather should be publically punished (stresses). When they are doing such a heinous crime then they have no right to be called as 'human beings' at all." $_{(Kcon+TB)}$	Deontological

(continued)

Table 7.1 (continued)

S.No.	Topic	Ethical issues	Case's SMK/opinions/beliefs/metaphors	Related ethical theory
3.	Contraception	3.1. Religious ban on contraceptives 3.2. Contraceptives can be abortifacient 3.3. Fringing right to birth and right to life 3.4. Conceptus as a potential human being	"The use should only be under medical observation and doctor's prescription, then it is fine." (Karg) "Sometimes certain arguments based on any religious belief, or false notion lead to objection in their use."	Deontological
			"I think that a person needs to decide for oneself that what decision should he/she make and should also think about the society. In today's time one should be very clear that how much resources can one invest, for instance, having two to three kids and not being able to take care of them would be again problematic. So, being a human being one should be sentient toward all these aspects and not like other animals." (TB+SMK)	Consequentialist/ utilitarian
			"Exceptions can happen, for instance in one of the cases, even after using Cu-T, fertilization did occur which may happen due to improper placing of Cu-T, etc." (Karg+SMK) "These pills are basically steroidal in nature, include a combination of sex hormones and can have varied effects on different people, sometimes they even interfere with the hormonal system of the body." (SMK)	Consequentialist
4.	MTP or abortion	4.1. Right to birth and right to life 4.2. Foetus as a potential adult 4.3. Principle of sentience 4.4. Instances where MTP is ethically appropriate	"There are certain cases where abortion becomes permissible as per the law, such as the cases where fetus is suffering from a serious disorder, may be physical or mental, or in cases of sexual assault, etc." (SMK+Karg)	Women's reproductive right

(continued)

Table 7.1 (continued)

S.No.	Topic	Ethical issues	Case's SMK/opinions/beliefs/metaphors	Related ethical theory
5.	In vitro fertilization	5.1. Safety of the technique in terms of mother's health 5.2. Wastage of embryos 5.3. Estranged relationships between child–parent, child–sibling, child–society, etc.	"For me, I definitely think that there is a better option, there are so many places from where you can adopt a child, so why not adopt a child?" (TB) "All these techniques have to be taken with a pinch of salt." (TB)	Consequentialist
6.	GM crops	6.1. Do we really need GM? 6.2. After-effects of GM on human health and environment 6.3. Accessibility and means to buy GM crops/seeds 6.4. Is organic food a better option? 6.5. Gene patents and biopiracy	"GM crops have not been that successful even in other countries." (SMK) "It has largely affected the socio-economic conditions of our farmers as well. Without understanding the complexity of the issue and without knowing what harm they may cause, these GM crops have been introduced into the market." (TB+Karg) "It is not just one crop that they have lost but also the fertility of the soil." (TB+Karg) "We don't have a separate space or land as such where we can do or perform any experiment, and then the weather conditions, the soil conditions are different. What may be good for U.S or other countries may not be good for India." (SMK+TB+Karg)	Consequentialist
			"I think organic farming is a better choice. But, everyone is not aware about these and the organic crops are slightly more expensive for the common man to purchase, but if they can be made cheaper and easily available, then I think we will not remember about GM crops." (SMK+TB+Karg) "Means that whatever is there in our heritage and culture, they just wanted to patent that." (SMK)	Principle of justice (deontological)

(continued)

Table 7.1 (continued)

S.No.	Topic	Ethical issues	Case's SMK/opinions/beliefs/metaphors	Related ethical theory
7.	Human genomic project	7.1. Principle of privacy and confidentiality of genetic information 7.2. Playing with nature's diversity and gene pool 7.3. Issue of risking human life 7.4. Issue of informed consent 7.5. Gene patent	"The intention behind the project was good. They did not disclose the identity of the person whose genome was taken for the study." (SMK+KArg)	Right to privacy and informed consent
			"Making genetically engineered babies is highly unethical, what are we trying to do? We are trying to behave as God. We are trying to play with the nature and nature will not leave us. Nature will set it right." (SMK+TB+Karg)	Rights of the foetus (deontological)
8.	Stem-cell therapy and human cloning	8.1. Issue of potentiality 8.2. Issue of human rights	"Stem cell therapy, one good thing about it is that we are not involving any other organism over here. These days, people like some famous celebrities are preserving the umbilical cords of their babies as they are a rich source of stem cells and knowing well the potency of these stem cells. So, stem cells are highly useful as they are undifferentiated and can be used for organ transplantation." (SMK+KArg)	Utilitarian
			"It can happen in the case of female foetus, where some parents can sacrifice or do away with their female child even at the embryo stage, which goes against the basic human rights of a girl child and should strictly be prohibited." (TB+SMK+Kcon+Karg)	Right to life (deontological)

(continued)

Table 7.1 (continued)

S.No.	Topic	Ethical issues	Case's SMK/opinions/beliefs/metaphors	Related ethical theory
9.	Genetically modified organisms	9.1 Misuse of GMOs as bioweapons 9.2 Limitation of principle of utilitarianism 9.3 Case-by-case analysis as the best ethical approach 9.4 Pain and suffering experienced by the concerned animals 9.5 "Playing God" and tinkering with the blueprints of life forms 9.6 Intrinsic respect and consideration for other life forms 9.7 'Transgenesis' leading towards breaching of species barrier	"We are creating something for our benefit, and little do these people and scientists who are involved in the research think about the Biosafety or even the Bioethics that should be followed." (SMK+TB+Kcon)	Ethic of safety (deontological)

Source In-depth interview transcripts of Meena

Table 7.2 Argument analysis based on Toulmin's method (average score = 2.6)

S.No.	Argument topic	Data	Warrant	Qualifier	Backing	Rebuttal	Claim
1.	Animal experimentation (Level = 3)	Animal experimentation does have an impact in granting skill of surgery	Removing all the dissections and animal experiments from the course structure is a bit ridiculous	Definitely killing the animals for fun is not the case here	Which is in the right of human benefit	Just because it is an animal we cannot take it for granted, we have to follow certain guidelines Use of animals can be reduced wherever it can be	Animal experimentation is very important
2.	Amniocentesis and sex-determination (Level = 3)	Because giving birth to an abnormal child will not only be risky for the foetus, but for the mother, sibling and for the society at large	We do not want that 'bad genetic material' to circulate in the society		Consequentialist	(Sex-determination) Strict enforcement of the law should be there. Such people practicing such a cruel act should be punished	Amniocentesis is equally important if the mother's age is above 35 or so
3.	Contraception (Level = 4)	In today's time, it should be very clear that how much money do you need to invest, just by having two to three kids and not able to take care of them		The use should only be under medical observation and doctor's prescription	Being a human being one should be sentient towards all these aspects	These pills are basically steroidal in nature, which are the sex hormones and can have varied effects on different people, sometimes they even interfere with the hormonal system of the body	It is fine
4.	MTP (Level = 1)	Such as the cases where the foetus is suffering from a serious disorder, may be physical or mental, or in cases of sexual assault, etc.					There are certain cases where abortion becomes permissible as per the law stated

(continued)

Table 7.2 (continued)

S.No.	Argument topic	Data	Warrant	Qualifier	Backing	Rebuttal	Claim
5.	IVF (Level = 1)	Surrogacy is another issue	All these techniques have to be taken with a pinch of salt		Consequentialist		Why not adopt a child?
6.	GM crops (Level = 3)	It has largely affected the socio-economic conditions of our farmers as well. Without understanding the complexity of the issue and without knowing what harm it may cause, these GM crops have been introduced into the market	It is not just one crop that they have lost but also the fertility of the soil	If they can be made cheaper and easily available, then I think we will not remember about GM crops	consequentialist	But everyone is not aware of these and the organic crops are slightly more expensive for the common man to purchase	I think organic farming is a better choice
7.	Stem-cell therapy (Level = 3)	One good thing about it is that we are not involving any other organism over here. These days, people like some famous celebrities are preserving the umbilical cords of their babies as they are a rich source of stem cells			Utilitarian	Some parents can sacrifice or donate their female child right at the embryo stage, which will go against the basic human rights of a girl child and should strictly be prohibited	Stem cells are highly useful as they are undifferentiated and can be used for organ transplantation

Table 7.3 Meena's emotional metaphors

Planes	Emotional metaphors
Individual	**Related to subject matter of ethical issues in biology** **Amniocentesis** 'equal value' as well as status to the girl child Such people practicing such a 'cruel act' should be punished... rather should be publically punished (stresses) **IVF** They were "happy" that they will have their own child but on the other hand, they were also "worried" about the fact that what will happen if it resulted in three or four embryos together "All these techniques have to be taken with a pinch of salt." Every other day they were supposed to go through ultrasounds and other tests that were also very "painful" plus economically also "draining" **GM crops** "It is not just one crop that they have lost but also the fertility of the soil." "That soil was the only thing they had for their survival (sympathetically)... and they could not pay back the loans, and had frustration and the Government agencies were not there for their help." "They felt like cheated, cheated by big companies, and also by the Government." "What may be good for U.S or other countries may not be good for India. Organic farming can be a wonderful way of growing crops" **GMOs** "I sometimes "fear" that where are we heading in the name of technology and Biotechnology? We are creating something for our benefit, and little do these people and scientists who are involved in the research think about the bio safety or even the Bioethics what should be followed." **HGP** We are trying to behave as a God. We are trying to play with the nature and nature will not leave us. Nature will set it right **Cloning** Human brain is always "greedy" **Related to teaching** I am a Jain person, no doubt about it, but I am a teacher in the class and my religion there is to teach Zoology to the best of my capabilities and if the syllabus requires that I've to do dissections then I would do it and it has nothing to do with my Jainism
Relational	They do understand it.... And they do appreciate Biotechnology on one end, but they are also equally concerned that how to ensure biosafety I try to pose simple questions in between the class to keep them awake as well as making them think, I am not seeking correct answers from them always
Sociopolitical	In today's time, it should be very clear that how much money do you need to invest, just by having two to three kids and not able to "take care" of them, so being a human being one should be "sentient" towards all these aspects and not like other animals

7.9 Notions About Students' Understanding, Difficulties and Alternative Conceptions with Respect to Ethical Issues in Biological Sciences

The present teacher is of the opinion that students do understand ethical issues and concerns related to recent Biotechnological advances, but lack in their ability to take sound decisions.

> They do understand it… and they do appreciate Biotechnology on one end, but they are also equally concerned that how to ensure biosafety. As witnessed in the class, the students are very much worried about what will be the consequences if the modified gene enters their system and the environment. They are also trying to see the other side of the coin that is the implementation of safety measures. $_{(KSU)}$

Thus, from the teacher's statement it is quite evident that students are aware of both the pros and cons of a particular technology, and regard the issue of biosafety to be implemented stringently so as to secure the safety and regulation of biotechnology. However, from the answer scripts that she evaluated as well as from her feedback based on students' responses in the classroom, she feels that students understand the concepts well, but it is often difficult for them to articulate things on paper. One important difficulty encountered by her students as pointed out by her was,

> Sometimes, when they are not able to correlate the things, if suppose we talk only theoretical without actually showing it to them, then it is a problematic thing $_{(KSU)}$. I try to pose simple questions in between the class to keep them awake as well as making them think, I am not seeking correct answers from them always, and in today's class you must have observed that there was a variety of answers to such a simple question and I appreciate that. $_{(KSU+KTS)}$

From the above statement made by the teacher, it emerges that students find it difficult to relate to the theoretical concepts without being given any practical exposure, thus making the content appear abstract and alienated to them. Secondly, the teacher poses simple questions in the class so as to grab their attention and without expecting correct answers from them always. This is a characteristic virtue of a democratic teacher who allows and shares students' opinions in the class and students feel free to answer in diverse ways without fear of being rebuked. The teacher rates her students' understanding somewhere between the ethical dilemma level and ethical decision-making level of understanding, and says that students are aware of these issues but are right now not in a position to take decisions $_{(KSU)}$.

7.10 Discussion

The present teacher's PCK is majorly constituted by Subject Matter Knowledge (SMK), her Beliefs (TB) and knowledge of argumentation (KArg) with respect to ethical issues in Biological Sciences (refer to Table 7.4 and Fig. 7.1). Although, SMK and Karg are at times overlapping and aid in generating and sustaining an ethical discourse in the classroom. Karg also determines how SMK is being presented to the students to make it more comprehensible. Teacher's beliefs and understanding of some of the ethical issues also contribute to the knowledge structure and influence the direction that the classroom discussion on ethical issues takes. This can especially be seen in the excerpts from her classroom teaching where teacher's beliefs which are mostly anti-GM have influenced and majorly directed the classroom discussion in terms of the arguments being raised both by the teacher and her students. The major transactional strategy adopted in dealing with the ethical issues includes discussions with the students, where students are active participants and contribute to the discussion. But an important point to note here is that the discussion does not end with a resolution of the issue at hand, as the ethical arguments are just being stated by the teacher or the students. An in-depth investigation into the authenticity and reliability of those arguments is lacking in the present teacher's classroom. Neither, anyone takes into consideration the source of those arguments, i.e. whether the arguments are being lifted from newspaper, journals, web-source, scientific community, media, books or just by hearsay. The consensus viewpoint reached in most of the ethical issues is development of technology but with regulation and laws in favour of consumer safety and security. The main ethical theory that the teacher takes recourse to maximum number of times happens to be Utilitarian (or Consequentialist) and deontological as is evident when she says,

Table 7.4 Teacher's knowledge components constituting PCK

S.No.	Teacher's knowledge component (code)	Evidence (in classroom interaction, discussion, interview, questionnaires, lesson plans, reflective fastwrites)	Frequency
1.	Subject Matter Knowledge (SMK)	"The two systems, i.e., the plant system and animal system, have different set of enzymes. Although, it is a very good idea to begin with the edible vaccines, as one is getting here a dual benefit of filling up one's appetite as well as getting immune to a set of diseases."	31
2.	Teacher's Beliefs (TB)	"For me, I definitely think that there is a better option, there are so many places from where you can adopt a child, so why not adopt a child?"	12
3.	Knowledge of Transactional Strategies (KTS)	"Teacher should be open and allow presentation n questions, noting down, guide students and telling them think out of box, a middle view all the aspects should be shown to students, but if going in wrong direction or weird, appreciate the differing opinions, to analyze merits n demerits. Students will start appreciating."	8

(continued)

Table 7.4 (continued)

S.No.	Teacher's knowledge component (code)	Evidence (in classroom interaction, discussion, interview, questionnaires, lesson plans, reflective fastwrites)	Frequency
4.	Knowledge of the Curriculum (Kcur)	"I definitely feel that the syllabus can be redefined and use of animals can be reduced wherever it can be, for instance for extracting one particular tissue we are dissecting the whole animal, here pooling can be done, for instance if dissection for viewing the animal system is being done then at the same time if the other teacher is present, the tissues of the same animal can be used for making histology slides."	3
5.	Knowledge of Students' Understanding (KSU)	"Sometimes, when they are not able to correlate the things, if suppose we talk only theoretical without actually showing it to them, then it is a problematic thing."	4
6.	Knowledge of Context (Kcon)	"We are creating something for our benefit, and little do these people and scientists who are involved in the research think about the bio safety or even the Bioethics what should be followed."	7
7.	Knowledge of Assessment (KA)		0
8.	Knowledge of Ethical Theories/Role of Ethics (in general and in science) (KET)	"Ethics is not only about right and wrong but also emphasizes on objectivity, integrity, confidentiality, social responsibility, animal care etc."	6
9.	Orientation towards Science Teaching (OST)	"To increase knowledge about the various aspects of relationship between humans and other organisms with the natural environment and also to increase knowledge about implementation of biological factors into various technologies."	4
10.	Knowledge of Ethical Argumentation (Raising Arguments/ Initiating Argumentation/ Resolving) (KArg)	"Genetically modified, which means that a foreign gene has been added to them, do you think that this foreign gene when ingested by us could cause any side-effect on a person's health?" "How can we be so sure that if a transgene is entering our body, it will not cause any altered reaction?"	34
11.	Knowledge about Nature of Science (KNOS)	"Scientific knowledge is reliable and infallible and scientific method is a strict procedure of framing a problem statement, devising hypothesis, collecting evidences, experimentation and reaching inference."	10
PCK_{Total}			**119**

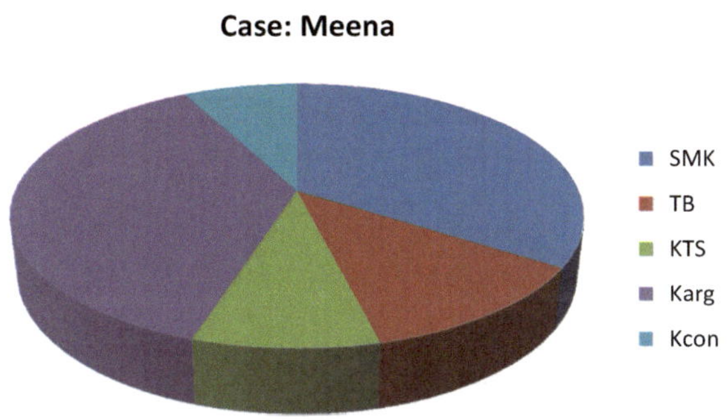

Fig. 7.1 Graphical representation of components of PCK

> Animal experimentation does have an impact in developing skill of surgery which is in the right of human benefit, which I think is more important. (Utilitarian)
>
> These pills are basically steroidal in nature, include a combination of sex hormones and can have varied effects on different people, sometimes they even interfere with the hormonal system of the body and interfere with it. (Consequentialist)
>
> There are certain cases where abortion becomes permissible as per the law stated, such as the cases where the fetus is suffering from a serious disorder, may be physical or mental, or in cases of sexual assault, etc. (Woman's Reproductive Right—Deontological).
>
> The present teacher relies on students' feedback and arguments as the main strategy for assessment of students' ethical understanding. However, no formal procedure for students' assessment is being carried out which depicts that not much importance is given to the teaching learning of ethical issues in Biological Sciences.

7.11 Conclusion

It can be ascertained from the above case study that the teacher can adopt an interdisciplinary approach to deal effectively with the ethical issues related to Biological Sciences rather than just restricting to the objective scientific approach. It is very well evident from the classroom discourses that scientific method is not the only method to reach a conclusion or decision with respect to a particular technology and its use, rather one often has to take recourse to ethico-moral inquiry to reach the 'right' decision. Besides, the choice of a 'right' decision or conclusion will be determined mainly by the universal ethical principles in case of a public decision and narrowly by the individual's own moral stance and values when it is about personal choice and stand on an issue. Thus, we always have divergent viewpoints on ethical issues related to Biotechnology which is one of the distinguishing attribute of ethical reasoning as ethical viewpoint should never be imposed upon by someone rather has to be acquired by an individual on her/his own. The present teacher succeeds in fostering such a divergent viewpoint among her students on these ethical issues, which she manages by adopting a variety of teaching–learning strategies. The transactional strategy includes interactive lectures, case studies followed by probing questions, asking open-ended questions, making use of emotional metaphors, presentations, videos, etc. Many Science teachers and educators world over have perceived the need for a paradigm shift towards an interdisciplinary science teaching (You, 2017). Although, the process is not as simplistic as seems to be, because the teachers often lack an awareness and understanding in the same respect. For nurturing such an understanding, teachers need to develop an interdisciplinary or multidisciplinary understanding of a particular concept and notice a meaningful pattern by drawing interlinkages. Negotiating ethical issues related to Biotechnology provides such a forum for facilitating higher order thinking among the learners and allows them to inquire, think critically, use deductive reasoning, analogies and synthetic thinking (Newell, 1998, 2002). The concept of 'interdisciplinarity' and interdisciplinary curriculum started gaining popularity in the mid-twentieth century (Klein, 1990; Vars, 1991). Tyler (1949) even regarded integration as important in establishing horizontal connections within the curriculum as well as necessary for an inquiry-oriented curriculum (Bloom, 1958). The idea of 'pedagogical progressivism' underscores the integration of disciplines and correlating with socially relevant themes that help in contextualizing the curriculum and catering to the needs and interests of the learners (Labaree, 2005). Similarly, ethical issues in Biotechnology provide a platform for integration of different disciplines and breaking the ideological barrier of looking only from a singular disciplinary lens. The present teacher provides many such instances and plots where the students can link diverse perspectives from social sciences, sociology, economics, sociopolitical context and ethical framework to comprehend the given issue. She uses a narration of case studies related to specific Biotechnological techniques and invites students' responses on the same. Thus, some components of critical pedagogy can be seen in her teaching learning of ethical issues in Biological Sciences.

References

Bloom, B. S. (1958). Ideas, problems, and methods of inquiry. *The Integration of Educational Experiences, 57,* 84–104.

Briscoe, C. (1991). The dynamic interactions among beliefs, role metaphors, and teaching practices: A case study of teacher change. *Science Education, 75*(2):185–199.

Clandinin, D. J. (1986). Classroom practice: Teachers' images in action. London: Falmer Press.

Klein, J. T. (1990). *Interdisciplinarity: History, theory, and practice*. Detroit, MI: Wayne State University Press.

Labaree, D. F. (2005). Progressivism, schools and schools of education: An American romance. *Paedagogica Historica, 41*(1–2), 275–288. https://doi.org/10.1080/0030923042000335583.

Lakoff, G., & Johnson, M. (1980). Metaphors we live by. Chicago: University of Chicago Press.

Newell, W. H. (Ed.). (1998). *Interdisciplinarity: Essays from the literature* (pp. 51–65). New York, NY: College Board.

Newell, W. H. (2002). Integrating the college curriculum. In J. T. Klein (Ed.), *Interdisciplinary education in K-12 and college* (pp. 119–137). New York: College Board Publications.

Tyler, R. W. (1949). *Basic principles of curriculum and instruction*. Chicago: University of Chicago Press.

Vars, G. F. (1991). Integrated curriculum in historical perspective. *Educational Leadership, 49*(2), 14–15.

You, H. S. (2017). Why teach science with an interdisciplinary approach—History, trends & conceptual frameworks. *Journal of Education and Learning, 6*(4), 66–77.

Chapter 8
An STSE (Science, Technology, Society, Environment) Approach for Teaching Ethics in Science—Case Narrative of an Undergrad Teacher

Abstract The present chapter presents a case study of an Undergrad Biology teacher to understand the different facets of his knowledge in the area of 'Ethical Issues in Biological Sciences'. These different knowledge domains include his idea about teaching, learning, pedagogy, curriculum, students' alternative conceptions, assessment, views on Nature of Science (NOS), etc., in the context of ethical issues in Biological Sciences. The present teacher often takes recourse to the STSE (Science, Technology, Society and Environment) approach and integrates the diverse perspectives for Negotiating ethical issues in Biological Sciences in the classroom. He used narrations, anecdotes, incorporated real-life experiences, arguments, emotional metaphors so as to build interlinkages between science, technology, society and environment as well as to generate ethical reasoning among students. Such a case study will also provide an insight into the categories of teachers' knowledge that are needed for addressing the ethical issues effectively in a Science classroom.

Keywords Ethical issues in biological sciences · STSE model · Ethical understanding · Pedagogy · Assessment

8.1 Introduction

Kunal teaches at one of the reputed colleges in Delhi region of India. He holds a Ph. D. degree in Zoology with specialization in 'Entomology' and has around twenty-one years of teaching experience at undergraduate level. His areas of interest include Genetics, Biotechnology and Molecular Biology. He has also undertaken some projects related to Biotechnology with his students and also motivates them toward learning the scientific processes involved therein. He prefers

A. Saxena, *Ethics in Science*, https://doi.org/10.1007/978-981-32-9009-9_8

teaching in a lecture cum discussion mode, involving student participation so as to make classroom an active platform for debate and discussion $_{(KTS)}$.[1] He feels that for fostering awareness about the newer technologies, more of student involvement as well as participation is required[2] $_{(OST+KTS)}$.

8.2 Idea About Teaching Biology at Undergraduate Level

The present teacher adheres to a set of broad range of objectives for teaching Biology at undergraduate level, such as,

> To generate awareness about the subject and its application, to develop love and passion for the subject and to enhance the analytical and observational skills.[3] $_{(OST)}$

About scientific literacy, he feels that it should also be one of the important goals of science education and should be imparted to each and every person.

> Understanding scientific principles, inculcating scientific temperament, linking the science for the welfare of the society in a holistic manner keeping in the view the environment, culture, ethics and values. $_{(OST)}$

Thus, the present teacher regards learning science not just to be limited within disciplinary boundaries rather should be interrelated and interdisciplinary and should always be directed towards the preservation of environment, culture, ethics and human values.

8.3 General Notions About Ethics in Science

The present teacher feels that sociology and ethics are as important as science and technology, and if a technology is entering the market, then the citizens need to be taught ethics of using it and associated values.[4] However, he did not seem to be clear with the idea of how ethics can be integrated within Science.

> "Consumerism, therefore we have to preserve moral values. Ethics is a set of rules or principles that needs to be followed in one's life for the welfare of humanity and environment."[5] $_{(KET)}$

---

[1]The present chapter involves the use of certain codes, e.g. KTS, OST, KNOS, etc., that refer to teacher's knowledge domains (for detailed reference see Appendix F).

[2]Source: Survey Proforma (Appendix C).

[3]Source: Survey Proforma (Appendix C).

[4]Source: Questionnaire (Appendix D).

[5]Source: Questionnaire (Appendix D).

The teacher possessed a normative idea about ethics and defined it as a set of rules or morals that people should adhere to. But how does ethics interferes with scientific understanding as well as in the process of decision-making with respect to any technology was not clearly explained.

8.4 Understanding about Nature of Science

The present teacher regards scientific knowledge as tentative, and modifiable in due course of time with better methodology. He looks at scientific method as a step-wise procedure involving looking into a problem, stating hypothesis, collecting observations, experimentation, testing the hypothesis and reaching a conclusion. He regards scientific processes as completely rational and objective especially when it comes to drawing inferences from a given data. An STS approach according to him is the only way to preserve the inherent value system of our society while conducting research in Science and Technology.

> STS is important to improve upon the life condition of man by applying scientific research without compromising the value system. $_{\text{(KNOS)}}$

In dealing with the multiple student views in the classroom, he feels that a teacher should open a discussion amongst the students and clear all their myths and misconceptions and explain by adopting a scientific approach. $_{\text{(KTS)}}$

8.5 Ways of Negotiating Ethical Issues in the Classroom

The major pedagogical approach adopted by the present teacher happens to be lecture cum discussion interspersed with some student arguments and inquiry $_{\text{(KTS)}}$. While dealing with the technique of transgenesis and creation of transgenic organisms, he presented the content before the class by using power-point presentation $_{\text{(KTS)}}$ and addressed some of the ethical issues associated with the technique by asking open-ended questions, problem-based questions, discussions with students (initiating ethical discourse in the classroom), and quoting some vignettes and examples $_{\text{(KTS)}}$. Some excerpts from his classroom interactions are worth mentioning in this regard.

In a classroom episode related to ‘transgenic organisms’, he raised a number of ethical concerns $_{\text{(KTS)}}$ related to the technology in order to generate sensitivity among students.

We are creating such mice who are suffering with Alzheimer's, and then we understand maybe some of the basic facts about the disease, symptoms, onset, molecular basis, diagnostics and also the pharmaceuticals which may be effective against the disease. (Utilitarian) $_{(KArg)}$ Is it really 'ethical' to create some of the diseases in mice which nature has protected them from? (principle of animal welfare and rights—Deontological) $_{(KArg)}$

It's not just about creating transgenic animals, but there are so many ethical issues attached to them, for instance, contamination of species integrity by using retroviral strains, are we answerable to the person who has consumed the transgenic poultry? Can we destroy any number of mouse embryos? (Preservation of species integrity—Deontologic ethic) $_{(KArg)}$

Sometimes, the teacher raised certain arguments and narrated some real or imaginary stories that pointed towards the ethical issues in sarcastically humorous ways $_{(KTS)}$. The following vignettes (refer to Vignettes 1, 2 and 3) taken from the present teacher's own narration can be cited as evidence for it.

Vignette 1: Organ Regeneration Using Transgenic Animals
"In 2008 probably, the total kidney transplants required in U.S was 80,000 however they could meet only 15,000 then what was the fate of the other 65,000, they were struggling and dying, so tissue transplant is always required in medical sciences, however it is a very very tricky and a very problematic procedure. Because it requires a suitable donor, and then there are some of the ethical issues associated with the commercialisation of the tissue graft, and not only commercialisation but an export of the tissue where the third world people are ready to sell their organs but still we don't have sufficient organs and donors. So, it was thought why not to use some livestock especially pigs whose size of organs is very much similar to human beings. But, obviously there will be difference in the structure, so we need to do some genetic modification in the pig, so that its organs can be transplanted in the human beings. Also the growth rate of pig is very fast and within two years we can get the desired organs, for e.g., heart, therefore if you need a heart transplant, then the pig's heart can be inserted, unfortunately you will be normal but with a pig's heart (laughingly)!!"

Vignette 2: Designer Babies and Cloning of organisms
2(a) "You can design a baby of your choice with the help of gene therapy, till now you could only design a dress of your choice. (Laughingly)"

2(b) "Bernard Shaw was the most intelligent persons of his time but was not at all good looking and Isadora Duncan who was one of the most beautiful and glamorous lady dancer, and once Isadora Duncan proposed to Bernard Shaw saying let's marry and have as beautiful child as I am and as intelligent as you are, to which Bernard Shaw said what if it happened otherwise?? (Laughingly)"

Vignette 3: Use of Animals in Transgenic Research
"HUMAN HEALTH, we should have a world full of healthy human beings, but are we really going healthy? And good health can only be achieved by understanding the disease and having 'good' medical advancement and at the same time if I say that you cannot do direct experiment on human beings as that would be much more unethical, so if you use a primate it is unethical, if you use a mice it is slightly lesser unethical, if fish then less, and with drosophila almost nil and is alright to be killed, so therefore, mouse we take as a model organism, advantage is that it is close to human being, but we every time say billi masi and never say chooha chacha!! (laughingly) emotionally, but actually chooha chacha is much more close to us than billi masi. So, chooha chacha is the appropriate model system, so, you got an idea? For good medical aid and even large quantity of food, and we have been selecting organisms with good traits and carrying out hybridisation, with the onset of human civilization, we have been destroying and disturbing the ecosystem, so, we have to live with it. Right now when there is such a huge population we need to have 'good' medical facilities and we also need to have large amount of food. So, where to get them from? Transgenic animal is the answer and we are NOT SELFISH as we are also giving disease resistance to the cattle also, so we are caring for them. So we need good drugs, good food and good environment, and right now these three-four logics are sufficient for us to go for transgenic animals (Utilitarian) but then we should not blindly follow, don't create 'designer babies' at least right now I say, may be after fifteen years I say, oh designer babies are very good and everyone should go for designer babies, we were fools that in our time the concept was not there that is why I find my children are like this. (Laughingly) Okey... So, it is always an ongoing process about the ethical issues, social issues, scientific issues, and technological issues and this debate is never-ending." (KArg)

The vignettes cited above helped in generating ethical arguments, for instance, in the first vignette on 'Organ regeneration using transgenesis' wherein the teacher mentioned about the pig heart being transplanted into the human and the implicit concern of tissue rejection and death due to foreign tissue not matching with the complement system of the recipient. Similarly in vignette 2, on 'designer babies and cloning', the teacher made some sarcastically humorous remarks by comparing the designing of babies with the designing of a dress, pointing towards the manoeuvre with which the concept is leveraged and promoted.

Box 8.1 Classroom Discourse on Transgenic Organisms

Teacher:	Although transgenics have huge applications but it is not free from flaws and controversies, the efficiency process is very lower, chances of contamination are there, then from ethical point of view you are inducing disease in the mice, do they also have a right to live and live a healthy life (Animal rights—Deontologic ethic)? $_{(KArg)}$ Are we turning cruel? We are creating and producing many traits which never existed in the natural population? We are creating new species which may turn out to be catastrophic. (Consequentialist) $_{(KArg)}$. Now, shall we go for transgenics? What are the major constraints?
Student 3:	They are unnatural.
Teacher:	Yes,… I completely agree with this point as already we are faced with so many different types of pollution and adding to it is the genetic pollution (Consequentialist) $_{(KArg)}$. So, this unnaturalness is the problem, what else?
Student 4:	We are disturbing the evolutionary process as well as harming the environment.
Teacher:	That's right, we are going against the process of evolution.
Student 5:	We are also providing the platform for new evolution.
Teacher:	Means that whatever is the natural course of evolution, we are deviating from it. $_{(KArg)}$
Student 6:	It may happen that some of the good or beneficial traits are also destroyed.
Teacher:	See,… "good" is a relative term, what is 'good' for me may not be 'good' for the species, for example, in the case of enviro pigs we may be creating environment friendly pigs which is good for us, but the pigs become leaner and suffer from arthritis which is bad for the species as such (Respect for an organism's life—Deontologic ethic and ethic of care). What is our role in this? We should not try to rule or master the universe, but live in harmony with all the animals, living and non-living entities (principle of justice and equality—Deontologic ethic). But, what we are doing, we are creating transgenic organisms as per our greed!! We are disturbing the balance of nature. What else? $_{(KArg)}$
Student 7:	Unconventionally undesired traits may be expressed.
Teacher:	Yes,… and if any of the organism with undesirable monstrous trait escapes, then what will happen? (Consequentialist) $_{(KArg)}$ What else? No one of you have ever thought from the animal's

point of view, their sufferings, nature has spared the animals from some of the diseases but out of our greed for research on some of the diseases we are inducing some of the diseases into them such as mouse for tuberculosis, alzeimer's etc. so, its UNETHICAL (Respect for Right to life of animals—Deontologic ethic) $_{\text{(KArg)}}$. I am thinking on all these points but everybody must have thought on these lines, and we need to justify our actions. Now, justify whether we should or should not go for transgenic research? $_{\text{(KTS)}}$

Student 8: Basically, when we talk about human survival.

The above classroom discourse (refer to Box 8.1) shows an active dialogue between the teacher and his students, involving equal turn-taking and opportunities to express one's opinion. The teacher did raise a number of ethical arguments with respect to 'transgenic organisms' which were explicit and clearly understood by the students $_{\text{(KTS)}}$. This was evident from the students' responses in the class; they were enthusiastic and generated some ethical arguments themselves. The questions being posed by the teacher were also simple and open-ended which resulted in a more interesting classroom interaction. He tried to limit the discussion on 'transgenic organisms' within the present context by linking it to the deplorable state of medical research where the use of transgenic organisms is inevitable. Although he seems to be friendlier with his learners and is receptive to all his student responses, but it was he who always validated them, structured and guided the discussion.

8.6 Understanding with Respect to the Ethical Issues in Biological Sciences

The in-depth interview sessions with the present teacher revealed his understanding and beliefs about some of the recent Biotechnologies. Some excerpts from the teacher's interview have been selected as they reflect some of the ethico-moral positions he takes with respect to various Biotechnologies along with their analysis based on ethical theories (refer to Table 8.1).

The present teacher's arguments are mostly based on 'Utilitarian Ethic', and regards human benefit at the top priority. According to his view, any technology, be it GM crops, GMOs, Gene therapy or stem-cell therapy should not in any way harm the human beings. The teacher is aware of the other side of most of the technologies, for instance, in the case of GM crops, the teacher is able to raise arguments which are both pro-GM and anti-GM.

Table 8.1 Analysis of teacher's arguments based on ethical theories

S. No.	Topic	Ethical issues	Case's SMK/opinions/beliefs/metaphors	Related ethical theory
1.	Animal experimentation	1.1. Use of animals in research and study 1.2. Pain and suffering to the animal 1.3. Wastage of animals	"A holistic approach can be developed only when they see the animal, they understand better, and have the hands-on experience then only they can relate with the significance of that animal." (TB) "Animal system is so complex that it cannot be taught unless they visualize it. And the students can really appreciate the anatomical details when they actually dissect the animal." (KTS+TB) "Importance of vision is there because what you see in the pictures, or books and when you see the actual animal, they are two different aspects." (TB)	Utilitarian
			"the very first thing we tell to the students before doing any dissection is that animal is not a toy to play, they are subjects for learning, and whatever the things that you have to see, focus on them only. And such experiments should not generate any kind of cruelty amongst the students." (KTS+TB)	Ethic of care and concern
			"'Pithing' an animal is a wrong practice and should not be done, at least I am against it." (TB) "Once an animal is open, it should be best utilized because OUR AIM IS LEARNING NOT KILLING." (TB)	3Rs—reduction, replacement and refinement—Deontologic ethic
2.	GM crops	2.1. Do we really need GM? 2.2. After-effects of GM on human health and environment 2.3. Accessibility and means to buy GM crops/seeds 2.4. Is organic food a better option? 2.5. Gene patents and biopiracy	"If we take just one approach for instance, green revolution or to say Integrated Pest Management (IPM), then it has been proved that it does not suffice the purpose. And a major share of crops is still damaged by insects and rodents." (KArg+TB) "I think this GM technology provides a very narrow approach and there is already an organism named diamond black moth, plutella, it has already developed resistance against Bt-Cotton. That's why I am saying that just one approach won't suffice, we need more holistic approaches." (KArg+TB)	
			"I think it can even generate some new and grave problems. The idea was to develop insect-resistant, drought-resistant, and better nutritional quality crops for the purpose of crop improvement." (SMK+KArg)	Consequentialist
			"But, at the same time we are creating an imbalance in nature. And also we are creating some of the social problems especially in the context of India. You know, if one of the farmer who's rich is growing Bt-cotton, then the farmer of the	Consequentialist Principle of beneficence

(continued)

Table 8.1 (continued)

S. No.	Topic	Ethical issues	Case's SMK/opinions/beliefs/metaphors	Related ethical theory
			adjacent field may not be so rich, and is growing an ordinary crop plant. Hence, all the insects will be diverted there, so the thinking is very narrow. Secondly, there is doubt that other organisms may also be affected." (SMK+KCon+KArg)	
			"The conditions in the lab are very different and in the field it is entirely different, so there are a lot of controversies, and unfortunately here there is involvement of multi-national companies who have spent millions of dollars in doing research so as to come up with this technology, so, they will definitely protect their products at any cost." (KCon+KArg) "Right now we are just aping the technology. However, this is the problem with all the hybrid seeds, as the seeds which we get from the F1 generation plants are usually of a mixed type and may not yield the same crop." (KArg+KCon+SMK)	Principle of Consequentialism
			"We also have to see the social context of a country, I think that's very important especially third world countries like India, the farmers are poor as far as self-sustaining and seeds are concerned. The patenting of GM seeds by foreign multinationals is also an issue of ethical concern." (KCon+KArg+SMK) "When we talk about GM crops, how many are actually there in the market? Sometimes biotechnology is even being referred to as 'science for labs', as safety is the main concern. You know when the golden rice were created and thrown into the market, they were not accepted. As the very first threat with GM technology is the retro-virus contamination." (KArg+KCon+SMK)	Issue of farmer's autonomy and Patent rights—Deontologic ethic
			"Organic farming is the need of the hour, but we should not forget that population is growing at a much faster and alarming pace, and we are currently facing a food and nutrition shortage. Why this golden rice were created, because vitamin A deficiency was a serious concern." (SMK+KArg+KCon)	Principle of safety and non-maleficence
3.	Genetically Modified Organisms (GMOs)	3.1. Misuse of GMOs as Bioweapons 3.2. Pain and suffering experienced by the concerned animal	"Technology is always a double-edged sword. And that needs to be seen, checked as well as regulated." (TB)	
			"If we look at it from the point of view of human health, then we need to study the disease first and test the drug. We cannot perform the experiments directly on the human beings. But, again there should be some ethical concerns; the significance of the experiment should be taken into consideration." (KArg)	Utilitarian

(continued)

Table 8.1 (continued)

S. No.	Topic	Ethical issues	Case's SMK/opinions/beliefs/metaphors	Related ethical theory
		3.3. "Playing God" and tinkering with the blueprints of life forms 3.4. Intrinsic respect and consideration for other life forms 3.5. "Transgenesis" leading towards breaching of species barrier	"The question before us is that to whom we shall give priority, human suffering or mice?" (KArg)	
			"Transgenic organisms right now are more in research, and then again we have to see the two sides, one is human welfare to which they are directly related, then at the same time, is the right of animals and their life. Many microorganisms are also widely used in the making of certain products." (KArg)	Utilitarian
			"These products are very much in market, and they are also safe and widely being used, such as pharmaceuticals, for example, Humulin (human insulin), growth hormone, etc. and they have really relieved some of the cases." (KArg+SMK+KCon) "However, there is one major limitation that all the human proteins cannot be synthesized by microorganisms, because they have a different system and we have a different system, moreover, there is a phenomenon of post-translational modification in humans." (KArg+SMK)	Utilitarian (principle of beneficence)
4.	Gene therapy	4.1. Playing with nature's diversity and gene pool 4.2. Issue of risking human life 4.3. Issue of informed consent	"Gene therapy…again it is an attempt to reduce the disease condition, but again it is at an infancy stage, still there are peaks and troughs, you are going to correct a particular gene and at the same time another gene is getting disturbed. That's why, the germ line gene therapy is already ruled out. Lot of things still need to be worked out." (KArg+SMK)	Principle of Consequentialism
5.	Stem-cell therapy and human cloning	5.1. Issue of potentiality 5.2. Issue of human rights	"As far as umbilical cord cells are concerned, there are no issues, but then comes the embryonic stem cells, which are derived from the embryos that are usually discarded or extra embryos produced during IVF procedure which raise a lot of ethical concerns as these embryos carry the potential of growing into a full-fledged human being." (SMK+KArg)	Utilitarian Right to Life

Source In-depth interviews with the case

> We should not forget that population is increasing at a much faster pace, and we are currently facing a food and nutrition shortage. Why this golden rice were created, because vitamin A deficiency was a serious concern. (pro-GM)

> I think this GM technology provides a very narrow approach and there is already an organism named diamond black moth, plutella, which has developed resistance against Bt-Cotton. (anti-GM)

> But, at the same time we are creating an imbalance in nature. And also we are creating some of the social problems especially in the context of India. You know, if one of the farmer who's rich is growing Bt-cotton, then the farmer of the adjacent field may not be so rich, and is growing an ordinary crop plant. Hence, all the insects will be diverted there, so the thinking is very narrow. Secondly, there is doubt that other organisms may also get affected. (anti-GM)

Thus, he seems to have a well-grounded and contextual understanding of technology. He is not overwhelmed by the advances and the benefits that the technology can offer, but is able to discern certain pitfalls and lacunas in the technology itself.

8.7 Sensitivity Towards Ethical Issues

The present teacher makes ample use of emotional metaphors[6] wherever required that reflect his sensitivity towards these ethical issues. This was evident both in his classroom interactions with his students as well as during in-depth interview sessions (refer to Table 8.2). Most of the emotional metaphors used by the teacher are expressed at an individual level while talking about the ethical issues during the in-depth interview sessions as well as classroom transactions.

8.8 Argument Analysis

The present teacher was able to make well-reasoned claims and arguments comprising all the components such as data, warrant, theory, qualifier and rebuttal (refer to Appendix H). The arguments usually have a strong rebuttal which reinforce his elaborative understanding vis-à-vis ethical issues related to biotechnology (refer to Table 8.3).

---

[6]Metaphorical language is said to be an extraordinarily powerful tool through which teachers express the meaning of their experiences in teaching and much of their pedagogical knowledge which teachers hold (e.g., see Briscoe, 1991; Clandinin, 1986). Lakoff & Johnson (1980) suggest that in order to understand one's thinking, and beliefs we should study the metaphors they use. Emotional Metaphors indicate how someone feels about events, objects or people in his/her life.

Table 8.2 Use of emotional metaphors

Planes	Emotional metaphors
Individual	**Animal experimentation** Only when they see the animal, they understand better, and have the hands-on experience then only they can 'relate' with the significance of that animal They are being told to handle it with 'care' as they have life. And such experiments should not generate any kind of 'cruelty' amongst the students **GM crops** (Empathetically) POVERTY, is there, which we can't understand, they just boil the rice and consume **GMOs** Technology is always a 'double-edged sword' Shall we leave them for 'suffering'? What will be the fate of such model organisms? Right now we are thinking from our point of view, but think about the 'poor animal' in which so many new genes have been expressed as a result of which they start showing arthritis You are inducing disease in the mice, do they also have a right to live and live a healthy life. Are we turning 'cruel'? We are creating transgenic organisms as per our 'greed' and not requirement but greed!! No one of you have ever thought from the animal's point of view, their sufferings We are 'NOT SELFISH' as we are also giving disease resistance to the cattle also, so we are 'caring' for them Gene therapy Can we play with the lives of human beings? We are in complete right to play with the lives of mice, primates, or drosophila sp.? **Teaching of the subject** They should develop 'love' to the type of studies that they are doing, 'passion' and understanding. So, that was the main approach and see the students were 'excited', they were 'enjoying' the class, participating I wish to 'sensitise' them about the limitations of a particular research I am taking up these issues out of my 'sheer interest' and 'care' towards them If I show only the 'green pastures', then they think that it is so 'wonderful', they don't know what are the limitations of this, what are the implications of this
Relational	
Sociopolitical	

8.9 Notion About the Curriculum at Undergraduate Stage with Respect to Ethical Issues in Biological Sciences

The present teacher strongly feels that these ethical issues should be appropriately addressed within the curriculum of Biological Sciences at undergraduate level for an informed understanding as well as better decision-making.

Table 8.3 Argument analysis based on Toulmin's method (average score = 4)

S. No.	Argument topic	Data	Warrant	Qualifier	Backing	Rebuttal	Claim
1.	Animal experimentation (Level = 4)	Animal organization is so complex that it cannot be taught unless they visualize it. And the students can really appreciate the architectural programme	Importance of vision is there because what you see in the pictures, or books and when you see the actual animal, they are two different aspects	Such experiments should not generate any kind of cruelty amongst the students	Animal is not a toy to play, they are subjects for learning	'Pithing' an animal is a wrong practice and should not be done, at least I am against it	When they see the animal, they understand better, and have the hands-on experience then only they can relate with the significance of that animal
2.	GM crops (Level = 5)	There is already an organism named diamond black moth, plutella, which has developed resistance against Bt-Cotton	At the same time, we are creating an imbalance in nature. And also we are creating some of the social problems especially in the context of India		Safety is the main concern	But we should not forget that population is increasing at a much faster pace, and we are currently facing a food and nutrition shortage Then, the conditions in the lab are very different and in the field it is entirely different, so still there are a lot of controversies, and unfortunately here there is an involvement of multinational companies who have spent millions of dollars in doing research	GM technology provides a very narrow approach Organic farming is the need of the hour

(continued)

Table 8.3 (continued)

S. No.	Argument topic	Data	Warrant	Qualifier	Backing	Rebuttal	Claim
3.	GMOs (Level = 5)	These products are very much in market, and they are also safe and widely being used, such as pharmaceuticals, example is Humulin (human insulin), growth hormone was there and they have really relieved some of the cases	If we look at it from the point of view of human health, then we need to study the disease first. We cannot perform the experiments directly on the human beings		Technology is always a double-edged sword. And that needs to be seen, checked as well as regulated	There is one major limitation that all the human proteins cannot be synthesized by microorganisms. Because, they have a different system and we have a different system, moreover, there is a phenomenon of post-translational modification in humans	we have to see the two waves, human welfare, to which they are directly related, then at the same time, is the right of animals
4.	Gene therapy (Level = 4)	Gene therapy…again it is an attempt, to reduce the disease condition	You are going to correct a particular gene and at the same time another gene is getting disturbed		Consequentialist	Germ line gene therapy is already ruled out. Lot of things still need to be worked out	It is at an infancy stage
5.	Stem-cell therapy (Level = 2)	There was a case, there was a billionaire who died in a plane crash, then who will be the heir to the property, and there were some leftover embryos	Whatever the embryos are being used are usually the leftovers of the IVF procedure, that ultimately lead to death		Utilitarian		As far as umbilical cord cells are concerned, there are no issues

> I think that all such issues should be taught to the students, then only they can understand and analyse the problem precisely. Just, teaching one aspect is not sufficient, if I show only the green pastures, then they think that it is so wonderful, they don't know what the limitations of this are, what are the implications of this $_{(KSU+KTS)}$. Although, Biotechnology has many promises, but we cannot just see from one angle, as biotechnology can be the biggest evil also, as bio-weapons can be created. So, a holistic approach should be adopted while dealing with such topics in the class. And ethical, legal and social issues should be taught. $_{(KCur+KTS+OST)}$

He feels that students should be sensitized towards the pros and cons of the Biotechnological procedures, the outcomes and risks involved, such as,

> The very first thing is …t..h..a..t… they should be told about the problem, about the biotechnological procedures that we are adopting, then they should be sensitised. Let them think and debate over that, because right now we cannot pack up with the Biotechnology. It should be adopted in a more appropriate manner. $_{(KTS+TB)}$

Thus, the teacher is of the view that students should also be made stakeholders in the decision-making process at least for themselves as well as for conducting research in the area, and should unhesitatingly debate and deliberate on the issues involved therein. However, he does regard time as one of the limiting factors that impede the kind of discussions required to address these issues in the classroom.

8.10 Discussion

An overall analysis of the data collected from classroom observations, questionnaires, in-depth interviews, informal conversations and general reflections indicates that the present teacher's Pedagogical Content Knowledge (PCK) vis-à-vis ethical issues mainly comprises of SMK, KArg, TB and KTS (refer to Table 8.4 and Fig. 8.1). He makes an effort to innovate the teaching–learning strategies in the classroom for addressing the ethical issues by posing open-ended questions to the students, problem-based learning that check the decision-making abilities of the learners, narrating case stories and initiating discussions based on them. He holds a sound understanding of the importance of context in learning which is evident from a number of instances during the classroom discussion wherein he based the arguments on the present-day context, such as,

> But, at the same time we are creating an imbalance in nature. And also we are creating some of the social problems especially in the context of India. You know, if one of the farmers who's rich is growing Bt-cotton, then the farmer of the adjacent field may not be so rich, and is growing an ordinary crop plant. Hence, all the insects will be diverting there, so the thinking is very narrow. Secondly, there is doubt that other organisms are also affected.

However, the teacher scored a minimum on KSU, KCur and KET, and almost a nil KA with respect to ethical issues. KSU or knowledge about students' understanding could be augmented by having frequent discussions with the students and gauging their understanding of these issues. The present teacher demands an

Table 8.4 Teacher's knowledge components constituting PCK

S. No.	Teacher's knowledge component (code)	Evidence (in classroom interaction, discussion, interview, questionnaires, lesson plans, reflective fastwrites)	Frequency
1.	Subject Matter Knowledge (SMK)	"A person is suffering from diabetes, then the first line of treatment offered is the restriction on the sugar intake. The second line of treatment would be giving insulin from an exogenous source, such that the donor would be a cow, goat or any other unrelated species. But, I want to be more specific, the third approach would be to give the missing gene product by using recombinant DNA technology"	40
2.	Teacher's Beliefs (TB)	"A holistic approach they will develop only when they see the animal, they understand better, and have the hands-on experience then only they can relate with the significance of that animal."	10
3.	Knowledge of Transactional Strategies (KTS)	Sometimes, the teacher raised certain arguments and narrated some case stories that pointed towards the ethical issues but in a sarcastically humorous way	28
4.	Knowledge of the Curriculum (KCur)	"I think that all such issues be taught to the students, then only they can understand and analyse the problem precisely. Just, teaching one aspect is not sufficient, if I show only the green pastures, then they think that it is so wonderful, they don't know what the limitations of this are, what are the implications of this."	4
5.	Knowledge of Students' Understanding (KSU)	"They should be told about the problem, about the biotechnological procedures that we are adopting, then they should be sensitized. Let them think and debate over that, because right now we cannot pack up with the biotechnology."	5
6.	Knowledge of Context (KCon)	"Then, the conditions in the lab are very different and in the field it is entirely different, so there are a lot of controversies, and unfortunately here there is an involvement of multi-national companies who have spent millions of dollars in doing research so as to come up with this technology, so, they will definitely protect their products."	10

(continued)

Table 8.4 (continued)

S. No.	Teacher's knowledge component (code)	Evidence (in classroom interaction, discussion, interview, questionnaires, lesson plans, reflective fastwrites)	Frequency
7.	Knowledge of Assessment (KA)		0
8.	Knowledge of Ethical Theories/ Role of Ethics (in general and in science) (KET)	"Ethics is a set of rules or principles that needs to be followed in one's life for the welfare of humanity and environment."	2
9.	Orientation towards Science Teaching (OST)	"Understanding scientific principles, inculcating scientific temperament, linking the science for the welfare of the society in a holistic manner keeping in the view the environment, culture, ethics and values"	5
10.	Knowledge of Ethical Argumentation (raising arguments/ initiating argumentation/resolving) (KArg)	"If we look at it from the point of view of human health, then we need to study the disease first. We cannot perform the experiments directly on the human beings. But, again there should be some norms and ethics of doing experiments/tests on animals; the significance of experiment should be taken into consideration."	55
11.	Knowledge about Nature of Science (KNOS)	"STS is important to improve upon the life condition of man by applying scientific research without compromising the value system."	19
$\mathbf{PCK_{Total}}$			**178**

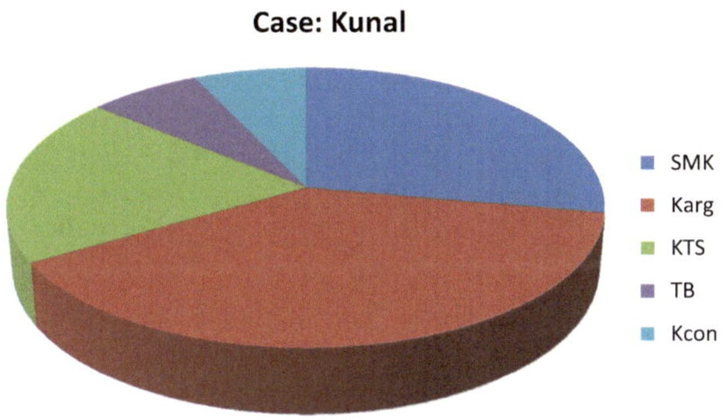

Fig. 8.1 Graphical representation of components of PCK

integrated science curriculum at undergraduate level for an appropriate addressal of ethical issues in order to enable students to view technology from a multidimensional perspective. The teacher takes into account STS/E approach to deal with the ethical issues in Biological Sciences. The use of this approach is quite evident in his classroom discourses especially when he narrates the stories and incidents depicting the impact of technology on the organismic life and the environment. According to him, technology can play awry and negatively impact human race and the environment, so it is important to understand it through an STS/E perspective that helps in establishing interlinkages between Science, Technology, Society and Environment. This is important especially in today's context when there has been a rise in industrialization and technology revolution in every sphere of endeavour leading to anthropocentrism and erosion of human values and ethics. Thus, following a STSE approach to curriculum transaction will help in reaffirming the place of core value system and ethics in scientific research.

8.11 Conclusion

The above case study of an Undergrad teacher depicts the role of different facets of his knowledge in the transaction of ethical issues in Biological Sciences. He is a strong adherent of STSE approach in Science Education. STSE education is an advanced form of STS education emphasizing upon the impact of Science and Technology advancements on Society as well as Environment. Thus, STSE education aims at understanding the interlinkages emerging at the interface of Science, Technology, Society and Environment (Chowdhury, 2016). In order to revive the faith in Science and mitigate its deleterious impact on society and environment, there is a need for the integration of ethical principles in Science. One of the earliest mentions about STSE education appeared in the late twentieth century in a paper written by Gallagher (1971),

> For future citizens in a democratic society, understanding the interrelationships of science, technology and society may be as important as understanding the concepts and processes of science (p. 337).

Ziman (1994) regards Science and technology as deeply embedded within the social, cultural, political and economic backdrop rather than considered as self-contained activities. This whole dimension is missing from the conventional science education. The sociocultural approach in STSE also helps in making Science accessible to all the learners (Costa, 1999; Jegede & Aikenhead, 1999) so that all the learners can equally participate in construction of knowledge. Socio-Scientific Issues (SSI) approach also evolved as a reform in Science education which shares similar principles, pedagogies and vision as STSE Education besides the fact that the focus of SSI is more on the psychological, epistemological growth and development of virtue in the child (Zeidler, Sadler, Simmons, & Howe, 2005, p. 358). Research has shown that students become more interested in learning Science when they are engaged in some authentic tasks and discussions that are relevant to their daily lives (Aikenhead, 2006; Ratcliffe & Grace, 2003). Also, such an approach helps in developing understanding of Nature of Science (NOS) and addressing citizenship education (Sadler, Barab, & Scott, 2007).

In the present case, the teacher tries to build the interlinkages between science, technology, society and environment by citing examples and instances from the application of technology in the society. He is successful in generating a dialogue and debate around the Bioethical issues in his classroom by giving ample space to his learners to express their opinion. He is also able to sensitise his learners about ethical issues by making use of certain emotional metaphors that do impact the teaching learning process (Zembylas, 2005). What is lacking is the assessment of ethical understanding of the learners on these issues, and reaching a particular stance vis-à-vis any technology. The students have only reached till the level of ethical dilemma but are not able to make sound decisions on a particular Biotechnology. This can be facilitated by engaging students in some projects that aim at honing their critical thinking and decision-making skills. These projects can be taken by the students themselves or with the assistance of the teacher. The teacher should assess these projects at regular intervals and grade the students on their level of ethical understanding.

References

Aikenhead, G. (2006). *Science education for everyday life: Evidence-based practice*. New York: Teachers College Press.

Briscoe, C. (1991). The dynamic interactions among beliefs, role metaphors, and teaching practices: A case study of teacher change. *Science Education, 75*(2), 185–199.

Chowdhury, M. A. (2016). The integration of science-technology-society/ science-technology-society-environment and socio-scientific-issues for effective science education and science teaching. *Electronic Journal of Science Education, 20*(5), 19–38.

Clandinin, D. J. (1986). Classroom practice: Teachers' images in action. London: Falmer Press.

Costa, V. (1999). How teacher and students study "all that matters" in high school chemistry. *International Journal of Science Education, 19*(9), 1005–1023.

Gallagher, J. J. (1971). A broader base for science teaching. *Science Education, 55*(3), 329–338.

Jegede, O. J., & Aikenhead, G. S. (1999). Transcending cultural borders: Implications for science teaching. *Research in Science & Technological Education, 17*(1), 45–66.

Lakoff, G., & Johnson, M. (1980). Metaphors we live by. Chicago: University of Chicago Press.

Ratcliffe, M., & Grace, M. (2003). *Science education for citizenship. Teaching socio-scientific issues*. Maidenhead: Open University Press.

Sadler, T. D., Barab, S. A., & Scott, B. (2007). What do students gain by engaging in socio-scientific inquiry? *Research in Science Education, 37*(4), 371–391.

Zeidler, D. L., Sadler, T. D., Simmons, M., & Howe, E. (2005). Beyond STS: A research-based framework for socioscientific issues education. *Science Education, 89*(3), 357–377.

Zembylas, M. (2005). Teaching with emotion: A postmodern enactment (p. 258). USA: Information Age Publishing, Inc.

Ziman, J. (1994). The rationale for STS is in the approach. In J. Solomon & G. Aikenhead (Eds.), *STS education: International perspectives on reform* (pp. 21–31). New York: Teachers College Press.

Chapter 9
Contextualizing Science for Ethical Understanding—Case Narrative of an Undergrad Teacher

Abstract The present chapter presents a case study of an Undergrad Biology teacher named *'Prakash'* in order to understand the different facets of his pedagogical content knowledge in the area of 'Ethical Issues in Biological Sciences'. The different knowledge domains include his idea about teaching, learning, pedagogy, curriculum, students' alternative conceptions, assessment of students' understanding, views on Nature of Science (NOS), etc. in the context of ethical issues in Biological Sciences. Transacting ethical issues within Biological Sciences in the classroom often requires contextualizing teaching and learning of Science and scientific concepts such that the students no more feel alienated with Science, technology and their applications in daily life. It is important for students to realize the everyday role of Science and technology and their stake in decision-making processes. For this to happen, the present teacher tries to make the teaching of Science as interactive and dramatic as possible, by including lived experiences, relevant and recent research, audio-visual aids, narratives and documentaries which kind of dissolves all hierarchies and power dynamics in the classroom and provides much needed freedom of thought and expression. Such a case study will also provide an insight into the categories of teachers' knowledge that need to be developed for addressing ethical issues in a Science classroom. The case study also provides exemplars of vignettes, classroom discourses, open-ended questions and other instances from the real classroom situations which may serve as a roadmap for teachers to address these issues in their classrooms.

Keywords High school · Biology teacher · Ethical issues in biological sciences · Context · Ethical understanding · Pedagogy · Assessment

9.1 Introduction to the Case

Prakash is a Senior Professor in the Zoology Department of a reputed college in Delhi. He has been teaching in the college for more than 30 years. He enjoys his work and is always receptive towards his students in helping them understand the

A. Saxena, *Ethics in Science*, https://doi.org/10.1007/978-981-32-9009-9_9

concepts related to the subject in the classroom as well as outside the classroom. He keeps himself abreast with all the recent research in the subject. He has a passion for the subject which is evident from the ways his classes are being conducted, which are more like open debate forums, where knowledge is not given rather co-constructed by the learners in the classroom. He likes to interact with his students, and there are no signs of any power relationship between the two, which is one of the prerequisite factors for a democratic classroom. Since ethical issues require more discussions and debate in the classroom, such kind of freedom and liberty is required, which the present teacher wholeheartedly gives to his students. He uses a variety of teaching–learning approaches in order to make his classes more interesting and vibrant. These include lecture with intermittent discussions, student responses and questions, while the beauty of his lectures is that they are not limited to the curriculum or the content rather go beyond taking into account certain historical and philosophical dimensions of science and linking it to the students' context.[1]

9.2 Notions About Teaching Biology at Undergraduate Level

The present teacher is of the view that students should not be burdened with too many factual details rather should be fostered to become well informed, able to rationalize and become 'good' human beings$_{(OST)}$.[2] His objective of teaching science is to generate the passion and interest for sciences at the younger level, and includes a day-to-day understanding about the life's processes and thinking rationally and logically $_{(OST)}$.

> The interest factor of the subject is particularly high among the students, as they want to know about themselves and about human society in general, certain sociological factors, so as to know about why we are and who we are.[3]

The vision for science education that the present teacher holds is adhered to the broader notions of scientific literacy which was evident from his following statement:

> The eventual goal of science education is to produce individuals capable of understanding and evaluating information that is, or claims to be, scientific in nature and of making decisions that incorporate that information appropriately and, furthermore, to produce a sufficient number and diversity of skilled and motivated future scientists, engineers, and other science-based professionals. Science education in Universities should build upon what has already been dealt with at school level - to develop critical basic knowledge and

[1]Source: Author.

[2]The present chapter involves the use of certain codes, e.g. KTS, OST, KNOS, etc. that refer to teacher's knowledge domains (for detailed reference see Appendix F).

[3]Source: Questionnaire (Appendix D).

> basic skills, interests, and habits of mind (e.g., to critically analyse and ask questions) which should help the student to apply their knowledge of science to personal and societal issues. (OST)

Thus, from these broader goals, the teacher aims at developing a critical understanding amongst the students in relation to science and society issues.

9.3 General Notions About Ethics in Science

The present teacher believes in individual ethics rather than a universal code of ethics, as ethics according to him is not a scientific subject, meaning that it has more to do with perceptions about morality $_{(KET)}$. Regarding the inclusion of ethical issues within the undergraduate curriculum of Biology, he said,

> All such ethical issues and the ethics involved therein needs to be discussed as we are living in a society, and we are bound by certain rules and regulations. Most of these laws are self-imposed, as human beings make these laws, and nobody is born to follow these laws $_{(KET)}$. Say, for instance women safety, certain kinds of attitudes need to be checked. No one is born with morals, but one has to learn them over the course of time. And with the learning of Biology we can know ourselves and others even better and that can help in changing and modulating attitudes. And I am completely in the favour of individual thought and freedom of expression, but if one's activity is going to affect another person's freedom then we need to think again, and a line needs to be drawn.[4] $_{(KET)}$

Thus, it is evident that the teacher holds a normative notion about ethics and favours a deontological ethic, when it comes to one's behaviour in the society and towards fellow beings, be it humans, animals, plants, environment, etc. However, he does feel that no one is born with a set of morals or rules of conduct, rather imbibes them over a course of time. This also means that morals or ethics cannot be taught to a person, and one's individuality and thought of expression need to be respected until it transgresses the freedom of another person.

9.4 Understanding of Nature of Science

The present teacher exhibits an above average level of understanding about Nature of Science (NOS) which cannot be just attributed to a heightened awareness rather a deeper engagement and a constant reflection over the issues associated with the nature of science such as the values a scientist should hold, method of science, nature of science as subjective, value-laden, and dynamic, role of a teacher in dealing with the ethico-moral dilemmas in Biotechnology and so on. His perspective about nature of science is majorly drawn from the historical and philosophical underpinnings of sciences, and he regards 'history of science' as a pivotal

[4]Source: Questionnaire (Appendix D).

component of science, both content-wise and helps in understanding scientific concepts, theories and laws in a better manner $_{(KNOS)}$.

> It definitely does! The history of science is full of great works and theories that have marked a turning point in the development of a branch of knowledge. As an example one can take the example of one of the most radical theory in Biology….that of the Theory of Evolution. Enunciation of the theory of Evolution brought about a change in the way all natural phenomena have come to be looked upon and at present the situation is such that "Nothing in Biology makes sense except in the light of Evolution". The history of science is peppered with numerous such instances where some basic principles and theories laid down the pathway for all subsequent developments in our understanding of the subject. A number of seminal works have laid down a new theoretical frame of reference or a new systematization of the known facts. From thence new specializations and even completely new disciplines have sprouted. In my view, a general exposure of the student to the historical background and the thought processes that led to advances through experimentation in her chosen discipline of science is a critical necessity for a complete and holistic understanding of the subject and for formulating further advances.

He regards that sciences and scientific temper have the ability to distinguish between different realities, such as epistemic, empirical, cultural, religious, etc., and provides a valid argument for this as follows:

> Since a "scientific temper" describes an attitude which involves the application of logic and the avoidance of bias and preconceived notions, anyone possessing it would question and analyse religious and cultural practices which are dependent on faith or practice. Empirical and observational situations would have the upper hand in anyone who has been exposed to and taught the scientific principles. Thus, science education is of critical importance in the growth of an individual and, by association, the growth of the nation.

The teacher agrees with the specific scientific, philosophical, intellectual (including mathematical), religious, technological, economic and cultural (including ethical and artistic) influences over scientists and science, especially when he said,

> The human mind is plastic as well as rigid. I am using these two values, which are opposite to each other, but together, advisedly. Science teaching tries to increase the plasticity….it teaches us not to be dogmatic and rigid. However, the natural resistance to change still persists in many people (in spite of exposure to scientific principles). Thus the specific scientific, philosophical, intellectual (including mathematical), religious, technological, economic and cultural (including ethical and artistic) circumstances in which scientists find themselves and work in, could colour the thinking processes and influence them. For example, there are quite a few scientists who would not give a second thought about visiting a temple or a god man after spending fruitful hours in their labs.

The teacher regards scientific knowledge in terms of scientific laws and principles as universal in nature and gives the examples of Newton's law of Gravitation, Newton's three laws of Classical Mechanics, Kepler's Laws on the Rotation of Planets, laws of thermodynamics and conservation of energy, etc. and says that these laws are universally applicable independent of space and time, although these physical dimensions are subjected to relativistic changes as assessed in the theory of relativity.

In dealing with varied students' opinions on an issue of ethical relevance, he feels that the role of the teacher should be to clearly distinguish between scientific and religious worldview.

> The teacher needs to be strictly scientific in her explanations. She needs to distinguish between religion and science and needs to point out that science is based on provable

experiments whereas religion is a matter of faith and belief, without an experimental verification possible of the views expressed in the religious books and scriptures. Thus, the present teacher tries to distinguish Science from other non-scientific claims by applying scientific method and processes. Also, the discipline of Science enables a person to become more rational and logical in approaching a given problem or issue rather than being led by false claims and propaganda.

9.5 Ways of Negotiating Ethical Issues in Biological Sciences

The classroom of the present teacher is truly interactive and full of enthusiasm and passion creating a vibrant learning environment. He uses a variety of methods in transacting ethical issues, and the most common approach adopted was linking the content with the daily life of the learner, making it appear more *contextual*. He does this by raising arguments (of all kinds, such as social, political, cultural, scientific as well as ethical arguments related to a particular technology) and initiating a dialogue with his students, giving examples, using videos and documentaries. However, his explanations sometimes become too elaborate that they take the space for students' arguments and viewpoints.

In a classroom episode on 'Transgenic Organisms' (refer to Box 9.1), the teacher is the one who is asking questions, raising different arguments, giving evidences and in a way directing the course of discussion. The students are not raising any ethical issue or even a clarifying question, rather are taking the teacher's word as final. Prakash's explanations are extraordinarily copious touching multiple dimensions of the issue of GM crops. These include monopoly established by foreign seed companies over indigenous seed variety, resistance developed against Bt-gene and growth of superbugs, reluctance on the part of people towards consumption of Bt-crops, farmers entering a vicious debt cycle leading to increasing farmer suicides.

Prakash also narrated real case stories (refer to Vignette-1, 2 and 3) for dealing with the ethical issues related to a particular topic/technology. These helped in linking the content with the context of the learners, and they could relate very well with it.

Vignette 1: Case of filing a patent against a living organism

A Genetic Engineer modified a bacterium of pseudomonas genus such that it was capable of breaking down crude oil (now known as *Pseudomonas putida*). The experiment involved multiple plasmids being incorporated into a single bacterium that would confer the bacteria ability to breakdown different hydrocarbon components of the crude oil. Such a genetically modified bacteria would help in breaking down oil from oil spills at a much faster rate. After the discovery and experiment, the scientist applied for a patent for his discovery and work at the U.S Patent office. But, his request was rejected as living organisms cannot be patented, then he appealed to higher courts and the matter finally got transferred to the supreme court of U.S.A. and finally his claim was accepted and patent was granted. Now, this was a landmark decision, as it was for the first time that a living organism got patented (on June 16, 1980). What do you think is the significance of this decision to crown a patent on a living organism?

Vignette 2: Case story of a farmer in Canada who was being duped and sued by Foreign seed companies for growing their hybrid seeds
There was a farmer in Canada, growing some variety of mustard and found that his neighbours were growing a similar variety of crop but genetically modified and slowly he found that their seeds were blown to his farm area and when he grew his next crop it came out to be genetically modified. Monsanto, the American multinational seed company sued the farmer on the ground that he is growing their patented crops without their consent. A case was being filed against the poor farmer and he was taken to the court. The court ruled against the poor farmer, now what I am trying to get across to you is the complexity of this whole situation. And in time, what is going to happen is that under all these pressures, these small farmers will finally say, lets grow the stuff that Monsanto is sending and get rid of all the court cases. Now, this means to say that slowly there would be a domination of such companies.

Vignette 3: Synthesising huge quantities of a protein using transgenic organisms – Example: Spider-silk Protein The spider silk is actually one of the strongest fibres known, and the applicability of such a product is enormous, such as your bullet-proof vests (Utilitarian). $_{\text{(KArg)}}$ Now, where do you get this spider silk from? Initially, the manufacturers started setting up spider webs and then they multiplied these spiders and started collecting silk fibres from them. However, you can imagine the total amount of silk that you can obtain from them. So, it was not a very viable option. Then, it occurred in the mind of a scientist who was researching on them to find out the gene that was responsible for making that protein. And then six genes were isolated. Then, came another step, that you take those six genes and put them in another system. Like, what they did was that they inserted those genes into a goat which has around 70,000 genes in which this extra gene for producing spider silk was been incorporated. Now, the milk which this goat produced had the spider silk protein, and the only thing to be done was to isolate that protein and spin it so as to obtain miles and miles of that fiber $_{\text{(SMK + KCon + KTS)}}$.

The discussion that was initiated after this is noteworthy from an ethical point of view (refer to Box 9.1).

Box 9.1 Classroom Discourse on Transgenic Organisms and their Applications

Student 4: But, is that fair to obtain fiber from goat's milk?

Prakash: Now, let's watch a small documentary on spider silk- source: Science Nation- Got Silk? http://www.youtube.com/watch?v=ktgACq4zcAU

After watching the documentary, a classroom discussion was being initiated by the professor so as to debate on the ethical consideration vis-à-vis animal transgenic technology $_{\text{(KTS)}}$.

Prakash: What could be the impact of such genetic modification on Goats? The scientist in the documentary spoke about their well being and no altered health effects, but how far would you agree with him? (Consequentialist) $_{\text{(KArg)}}$ Similarly in some of the slaughter-houses, in order to increase the muscle mass, animals are being treated with a gene which is inserted into them that causes them to become so fat and bulky that they can't even stand properly. Do we have any right to trouble animals to such an extent? (Animal rights- Deontologic ethic) $_{\text{(KArg)}}$.

Student 5: Everything needs boundaries.

Prakash: Regulations are necessary, just as in any other case. Devote some thought to such a scenario. To what extent can we go? Is it morally/ethically permissible? Who's to decide the fate of these goats? (Principle of Justice—Deontologic ethic) $_{\text{(KArg)}}$.

Student 6: Aren't these goats affected by this extra protein?

The above classroom discourse raised some serious ethical issues about the transgenic organisms, such as certain altered effects of the transgene on the organism including the impact on body muscles of these organisms, inability to walk properly, inflicting pain and trouble to the animal and using animals as a means to an end.

In another class, after discussing with the students about the pros and cons of oral contraceptives, especially emergency contraceptives, the teacher tried to generate students' opinions by showing them a documentary on i-pill and its side effects. At the end, he raised some questions in order to sensitize the learners on the various ethical issues attached with emergency contraceptives.

- How does one give protection to young female population? (SMK + KArg)
- What is life and when does life start? {at the level of fertilization, implantation or in the third trimester of pregnancy} (SMK + KArg)
- Whether contraceptives should be distributed freely? (SMK + KArg)
- Whether such pills be taken or not? (SMK + KArg)
- To what extent should the Government go to educate the population? (SMK + KArg)
- There are conflicting interests of the companies that want to increase their profit without being concerned about the consumer. (SMK + KArg)

- Government is concerned about the increasing population. (SMK + KArg)
- Individuals prefer a drug that is safe. (SMK + KArg)
- Under these conflicting situations what should be the right path? We are not dealing with only informed class, but poorest of the poor and richest of the rich. (SMK + KArg + KTS)
- To what extent does our young generation know about the physiology of sex, when should sex education start and at what level? (KTS)

However, there was not much discussion over these issues due to the sensitivity component attached to them.

9.6 Understanding with Respect to the Ethical Issues in Biological Sciences

The present teacher seems to have a sound understanding with respect to the ethical issues in Biological Sciences. This was evident from the arguments generated by him during the in-depth interview sessions where some of the arguments reflected hidden ethical stances and reliance on a particular ethical theory (refer to Table 9.1).

After analysing the various arguments given by the teacher, it appears that he is basically in the favour of technological advancement. Some technologies such as GM crops and other GMOs according to him are in their nascent stage and that is why nothing much can be said or predicted about them.

> If by use of transgenic technology you are able to produce large sized animals, but if such animals are not able to reproduce, they are not fertile, and if they are harmless, that is they don't cause any harm when ingested by humans, then I don't see any harm in producing such kinds of animals. (SMK + TB)

This shows that the present teacher supports a technology which is beneficial and harmless (principle of beneficence-consequential ethics). Another statement made by him in the context of human genomic project clearly reflects a utilitarian perspective,

> The applications of such a project wherein we are able to detect what all our genes constitute and what function do they perform are great, ranging from designer drugs to designer babies. (SMK)

Another such argument made by the teacher clearly showed that he gives priority to human life as against any other issue raised by a particular technology,

> All those people who argue that any kind of intervention during gestation is forbidden, if a foetus develops a defect in the heart, and then would these people still don't allow any medical intervention? Would that be right, because then the child will be malformed? (SMK + KArg)

An in-depth analysis of the teacher's arguments reveals that in some of the arguments he has a strong rebuttal whereas in others the rebuttal is weak. In the case

Table 9.1 Analysis of teacher's arguments based on ethical theories

S. No.	Topic	Ethical issues	Case's SMK/Opinions/Beliefs/Metaphors	Related ethical theory
1.	Animal experimentation	1.1. Use of animals in research and study 1.2. Pain and suffering to the animal 1.3. Wastage of animals	"My personal opinion is that we should definitely LIMIT it. The crucial point over here is that we are offering here a course in zoology, and if the dissections are eliminated then the student hardly has any direct contact with the live animal." (KArg+KCon) "The argument which comes is that why the students should touch live animals at all, as there are enough documentaries, photographs, etc. but, unless the student opens up an animal himself/herself, the FEEL of what it is like to handle an animal, that doesn't come in." (KArg) "And I think that it is very much necessary at the undergraduate level to discuss this issue, animal experimentation, the pros and cons of it, the ethical aspect. Why is it that some people are saying that we should not dissect animals? Why is it that the scientists in general and teachers at the University argue for animal experimentations otherwise the students won't get the feel of the subject?" (KCur+KTS+KArg)	Utilitarian Animal Rights—Deontologic ethic
2.	Amniocentesis	2.1. Sex determination 2.2. Female foeticide 2.3. Craving for the male child	"The technique by itself is useful and if done properly it is fairly safe. The misuse of the technique is basically….. in determining the sex of the unborn child." (SMK) "The difference between say India and America is that in America they generally don't bother whether the child is a male or a female. May be it is the social conditions that we live in are responsible for a preference toward the male child, but with education and awareness, that is going to recede in time." (TB) "I think first the Govt. needs to put in place a serious programme where these issues are brought to the notice of the public because the mindset has to change, as without the change in the mindset we are not going to solve this issue." (TB)	Utilitarian Basic human rights (Deontologic ethic)

(continued)

Table 9.1 (continued)

S. No.	Topic	Ethical issues	Case's SMK/Opinions/Beliefs/Metaphors	Related ethical theory
3.	Contraception	3.1. Religious ban on contraceptives 3.2. Contraceptives can be abortifacient 3.3. Fringing right to birth and right to life 3.4. Conceptus as a potential human being	"I don't think that God comes into this picture of usage of contraceptives. It is more of a necessity as far as our population is concerned, the extra number of people that a country can support and definitely it needs to be controlled." (SMK+TB) "MTP should not be used as a method of contraception as it is highly invasive and should be used only in emergency situations." (SMK)	Consequentialist
4.	In vitro fertilization	4.1. Safety of the technique in terms of mother's health 4.2. Wastage of embryos 4.3. Estranged relationships between child–parent, child–sibling, child–society, etc.	"They..are….successful, and over the years the percentage of success has been going up, it ranges around 40% at present which is pretty good and hopefully in the future, it might even go up more." (SMK) "You see…… there need not be a waste, as, such embryos are usually preserved or could be donated or could be used for research also." (SMK+TB) "IVF is generally an accepted procedure now and I don't think that anyone is seriously against IVF in principle. Initially there were arguments such as we are tampering with the nature's creation mechanism, but now it has been largely accepted as anyother medical procedure and if the people need it, they can access it." (SMK+TB+KArg+KCon)	Utilitarian Principle of beneficence
5.	GM crops	5.1. Do we really need GM? 5.2. After-effects of GM on human health and environment 5.3. Accessibility and means to buy GM crops/seeds. 5.4. Is organic food a better option? 5.5. Gene patents and biopiracy	"I don't fully support GM crops with the state of knowledge that we have at present. I do feel that a lot more research needs to be done, we need to be very sure when we are going to introduce these organisms into the nature because there's no way of going back once something goes wrong." (TB+KArg) "GM technology should not be our total focus; we should keep that as one of the methodologies which we can use for improving our condition. But, we should work on other technologies which are available now." (TB)	Principle of non-maleficence Principle of justice Deontologic ethic Principle of justice

(continued)

Table 9.1 (continued)

S. No.	Topic	Ethical issues	Case's SMK/Opinions/Beliefs/Metaphors	Related ethical theory
			"It's a GOOD concept, no doubt about it. Unfortunately there are some limitations, particularly of space. Since organic farming depends upon the natural means of cultivation such as use of natural or man-made fertilizers and manures, so it is very difficult to achieve that on a mass scale without the use of chemical fertilizers or pesticides so it would be difficult to go for organic farming on a big scale." (SMK+KArg) "But it is going to be costly and everyone cannot afford it which brings us again back to the question of inequality between the haves and have nots." (KArg) "If you are patenting a gene then you have to patent that organism also and that is something which I am against." "Even if you are doing some work on a gene, it should be an open access for everyone because it should not be used for merely enhancing the personal economic benefits. And should not be done that way as it is not right." (KArg+TB)	
6.	Genetically Modified Organisms (GMOs)	6.1. Misuse of GMOs as bioweapons 6.2. Pain and suffering experienced by the concerned animal 6.3. "Playing God" and tinkering with the blueprints of life forms 6.4. Intrinsic respect and consideration for other life forms 6.5. "Transgenesis" leading towards breaching of species barrier	"At present because we are just beginning our work on GMOs, we still don't know how the process works, producing transgenics is very inefficient at present. We're throwing away much more than we are actually creating out of it all." (SMK+TB) "If by use of transgenic technology you are able to produce large sized animals, but if such animals are not able to reproduce, they are not fertile, and if they are harmless, that is they don't cause any harm when ingested by humans, then I don't see any harm in producing such kinds of animals." (SMK+TB) "Under normal circumstances when a transgenic product is being ingested then it goes through the normal intestinal tract where everything gets broken up and after that there won't be any gene left as such. But, may be if that gene is taken up by some insect or virus that infects us, then it might come over to us." (SMK+TB)	Consequentialist Utilitarian Consequentialist

(continued)

Table 9.1 (continued)

S. No.	Topic	Ethical issues	Case's SMK/Opinions/Beliefs/Metaphors	Related ethical theory
7.	Human genomic project	7.1. Principle of privacy and confidentiality of genetic information 7.2. Playing with nature's diversity and gene pool 7.3. Issue of risking human life 7.4. Issue of informed consent 7.5. Gene patent	"The applications of such a project wherein we are able to detect what all our genes constitute and what function do they perform are great, ranging from designer drugs to designer babies." (SMK) "I think it is more natural for all human beings to get more money, more benefits, etc. maybe that is what is driving all such sort of research in biotechnology." (KArg+KCon) "I think it is now very clear that we are not our genes. So, definitely the genes are also affected by the environment that allows certain genes to turn on and off. And if that is the case then we can't say that genes are deterministic." (SMK) "As man has already been playing with nature since very long. And I really don't know whether it's a bad thing to play God." (TB)	Utilitarian Utilitarian
8.	Stem cell therapy and human cloning	8.1. Issue of potentiality 8.2. Issue of human rights	"I believe a person becomes a person only after birth or during the latter part of gestation." (TB) "All those people who argue that any kind of intervention during gestation is forbidden, if a foetus develops a defect in the heart, and then would these people still don't allow any medical intervention? Would that be right, because then the child will be malformed?" (SMK+KArg)	Consequentialist

Source In-depth interviews with the case

of GM crops, he is having two different but contrasting claims, unlike any other issue,

> It's a GOOD concept, no doubt about it.
>
> I don't fully support GM crops with the state of knowledge that we have at present.

This again indicates that the teacher is not able to take a firm stand on whether the technology should be promoted or not, but certainly favours research in this direction.

9.7 Sensitivity Towards Ethical Issues

The present teacher uses a number of emotional metaphors[5] in his classroom discussions with his students so as to create awareness and generate sensitivity towards these issues (refer to Table 9.2).

9.8 Argument Analysis

The argument analysis based on Toulmin's method of argument analysis (Toulmin, 1958, 2003) reveals that the teacher is able to frame good quality arguments (average score-3) supported with data, warrants and backing along with strong rebuttals (refer to Appendix G). The arguments reflect present teacher's ethical, social, political and scientific understanding on the issues related to Biotechnology (refer to Table 9.3).

In most of the cases, the teacher has a pragmatic approach for dealing with the ethical issues pertaining to recent technology. He is very much aware of the multiple arguments that surround each such technology but tends to take an anthropocentric side by favouring human benefit at the end.

9.9 Views Regarding Curriculum at Undergraduate Stage with Respect to Ethical Issues in Biological Sciences

The present teacher feels that the ethical issues have not received their due place in the curriculum of life sciences, Botany and Zoology at the undergraduate level and vies for it as according to him a rendezvous with these issues is important in this day and age of science and technology $_{\text{(KCur)}}$. He reflects upon the internal politics

[5]A 'Metaphor' helps in the transference of meaning from one situation to another and helps in a better understanding of a phenomenon. Emotional metaphors provide a useful insight about the ways teachers think, conceptualize about certain aspects of teaching and learning (Saban, Kocbeker, & Saban, 2007); about their learners (Saban, 2010) and their subject knowledge.

Table 9.2 Prakash's emotional metaphors

Planes	Emotional metaphors
Individual	*Animal experimentation* Unless the student opens up an animal himself/herself, the '*FEEL*' of what it is like to handle an animal, that doesn't come in *Amniocentesis* I think first the Govt. needs to put in place a '*serious*' programme where these issues are brought to the notice of the public because the *mindset has to change*, as without the change in the mindset we are not going to solve this issue *Contraceptives* It is more of a '*necessity*' as far as our population is concerned the extra number of people that a country can support and definitely needs to be controlled *GM crops* To what extent do monolithic or private organizations have a role in '*controlling*' human life? In countries like in Europe, there is public '*resistance*' as to the acceptance of these GMOs. And they say that it is composed of a lot of problems, it may lead to allergic reactions, etc. and even now nothing is proved this way or that way And in time what is going to happen is that under all these pressures, these small farmers will finally say, '*do hell with it*', and lets grow the stuff that Monsanto is sending and get rid of all the court cases. Now, which means to say that slowly there would be a '*domination*' of such companies what if a particular organization asks our farmers to plant and grow their seeds and then after the crop is harvested, the farmer is '*being forced*' to buy those seeds again from these organizations (Principle of Autonomy)? As a niche technique (organic farming) it is ok, but it is going to be costly and everyone cannot afford it which brings us again back to the question of inequality between the haves and have not's *GMOs* Do we have any right to '*trouble*' animals to such an extent? Teaching of the subject I am completely in the favor of individual thought and '*freedom*' of expression
Relational	
Sociopolitical	Overall I think that there is *interest* in these matters, unfortunately you know the problem is that the present system of education….. ummm….. these things are *not institutionalized*

that lead to the removal of ethical issues from the syllabus and offers some novel ideas for integrating ethics within the present curriculum, such as,

> I think that would come in only if we include a course on 'animal ethics'. And it is actually a requirement $_{\text{(KCur)}}$. And this depends upon a whole lot of factors, the question is that who exactly is removing the ethical component and if it was there in the first place and later got removed, then who's responsible? Unfortunately as the situation is now, it's in a state of flux, the reason is that the whole semester syllabus has been formulated in such a hurry that they have not looked at the time availability, the amount of work an individual teacher has to do to complete the course, so all those aspects. Now, what has happened is that the

Table 9.3 Argument analysis based on Toulmin's method (Average score = 3.1)

S. No.	Argument topic	Data	Warrant	Qualifier	Backing	Rebuttal	Claim
1.	Animal experimentation (Level = 3)	If the dissections are eliminated then the student hardly has any direct contact with the live animal	Unless the student opens up an animal himself/herself, the FEEL of what it is like to handle an animal, that does not come in		Utilitarian and Deontologic ethic	Why the students should touch live animals at all, as there are enough documentaries, photographs, etc.	We should definitely LIMIT it I think that it is very much necessary at the undergraduate level to discuss this issue, animal experimentation, the pros and cons of it, the ethical aspect
2.	Amniocentesis (Level = 4)	May be it is the social conditions that we live in are responsible for a preference towards the male child, but with education and awareness, that is going to recede in time	The misuse of the technique is basically….. in determining the sex of the unborn child		The Govt. needs to put in place a serious programme where these issues are brought to the notice of the general public	The technique by itself is useful and if done properly it is fairly safe	The mindset has to change, as without the change in the mindset we are not going to solve this issue
3.	Contraception (Level = 2)	Extra number of people that a country can support and definitely needs to be controlled					It is more of a necessity as far as our population is concerned
4.	IVF (Level = 3)	They..are….successful, and over the years the percentage of success has been going up, it ranges around 40% at present which is pretty good	There need not be a waste as such embryos are usually preserved or could be donated or could be used for research also		Principle of beneficence	Initially, there were arguments such as we are tampering with the nature's creation mechanism	IVF is generally an accepted procedure now

(continued)

Table 9.3 (continued)

S. No.	Argument topic	Data	Warrant	Qualifier	Backing	Rebuttal	Claim
5.	GM Crops (Level = 4)	There are some limitations, particularly of space	A lot more research needs to be done, we need to be very sure when we are going to introduce these organisms into the nature because there's no way of going back once something goes wrong		Principle of non-maleficence	It would be difficult to go for organic farming on a big scale	It is a GOOD concept, no doubt about it I do not fully support GM crops with the state of knowledge that we have at present
6.	GMOs (Level = 3)	Under normal circumstances when a transgenic product is being ingested then it goes through the normal intestinal tract where everything gets broken up and after that there will not be any gene left as such		if they are harmless, that is they do not cause any harm when ingested by humans	Consequentialist	But, may be if that gene is taken up by some insect or virus that infects us, then it might come over to us	I do not see any harm in producing such kinds of animals
7.	HGP (Level = 3)	The applications of such a project wherein we are able to detect what all our genes constitute and what function do they perform are great, ranging from designer drugs to designer babies	I think it is more natural for all human beings to get more money, more benefits, etc. maybe that is what is driving all such sort of research in biotechnology		Utilitarian	We cannot say that genes are deterministic	As man has already been playing with nature since very long. And I really do not know whether it is a bad thing to play God

teachers are sort of getting together; taking some adhoc decisions such as lets reduce this component, let us remove that component and deal with the mechanics of the subject because as per them this would be the only thing which is needed. Whereas ethics is something…… not that important and that's the reason these ethical issues have been removed from the syllabus $_{(KCur)}$. I personally feel that there should be one complete unit on Bioethics, as it is very necessary in the present day context. We need to sensitise not just our students but also our teachers so as to understand the inter-relationships that we share with the animal world and our environment. $_{(KCur)}$

9.10 Teacher's Beliefs Versus Students' Understanding on Ethical Issues

The students of the present teacher's class generated multiple arguments pertaining especially to GM crops,[6] and it seems that most of them are well informed about the issue and are also socio-ethically aware of the pros and cons that these technologies have to offer. They are sceptical in their arguments which clearly depicts that they are not ready to blindly follow whatever the science and technology has to offer rather are critical not only to the very idea and intentions behind a particular technology but also its impact on the human society and environment. They still stick to the natural means of growing the crops which is the organic way and regard it as safe and eco-friendly.

The present teacher holds a balanced opinion vis-à-vis GM technology and feels that instead of diverting all the resources towards the development of this technology we should also try and explore other options that are available, such as natural methods of growing crops or organic method, etc. He does not completely deny the need or efficiency of GM technology, but seems to be a little biased towards the other methods of crop improvement. The following statements made by him depict his stance related to the technology:

I don't fully support GM crops with the state of knowledge that we have at present. I do feel that a lot more research needs to be done, we need to be very sure when we are going to introduce these organisms into the nature because there's no way of going back once something goes wrong.

GM technology should not be our total focus; we should keep that as one of the methodologies which we can use for improving our condition. But, we should work on other technologies which are available now.

Thus, teacher's beliefs seem to be quite overlapping with his students' viewpoints which again indicate that teacher's beliefs do impact student understanding on a particular issue. His own assumptions about his students' understanding vis-à-vis these ethical issues are also very encouraging, as according to him some students are keen to understand and learn about these issues and find them interesting and challenging at the same time.

---

[6]For a detailed account of students' arguments and understanding on ethical issues, refer to Chap. 10.

> Some of them were very keenly interested in some of these ethical aspects. Also, when a teacher asks them that in a particular situation how would you react? They are ready to come out with their ideas. So, overall I think that there is interest in these matters, unfortunately you know the problem is that the present system of education….. ummm….. these things are not institutionalised. So, it's not in the syllabus as such. $_{(\text{KSU + KCur + KCon})}$

9.11 Discussion

The analysis reveals that one of the major components of the present teacher's PCK with respect to ethical issues is subject matter knowledge $_{(\text{SMK})}$ and knowledge of argumentation $_{(\text{KArg})}$ which can be considered a subcomponent of SMK. Knowledge of transactional strategies $_{(\text{KTS})}$, knowledge about the context $_{(\text{KCon})}$ and teacher's beliefs $_{(\text{TB})}$ constitute the other important components of his PCK (refer to Table 9.4 and Fig. 9.1). Besides, the teacher also holds a sound understanding of the nature of science and its various principles, which could also be one of the factors making his classes on ethical issues more effective. He uses real-life case stories, vignettes and examples from the daily life of the learners in order to link the content with the context and provide meaning to it. He gives complete freedom to his learners to argue and debate over an issue in the classroom. However, sometimes the explanations offered by him become so lengthy that they take over students' time for giving responses, and there is less time left for students' argumentation and discussion. He holds a general notion about the students' understanding about these ethical issues, and thinks that these issues interest students and they are aware of them.

The student notions and ideas with respect to ethical issues are well informed and often follow from the classroom discussion and deliberation on these issues. This indicates that classroom discussions and teacher's arguments do have an impact on students' understanding of these issues as well as argumentation. But, they are not able to take any stand with respect to GM technology claiming it on the inappropriate understanding of GM technology at present. They however feel informed as takers and buyers of the technology being science graduates.

Focussed group discussions with students revealed that there is definitely interest and concern in such matters of ethical relevance. And in the present teacher's case, it can be observed that how teacher's enthusiasm and sensitivity get transferred to his students who take an active part in classroom discussions vis-à-vis ethical issues in Biological Sciences. In order to generate such an informed understanding on such issues, the present teacher is well versed with not only the content knowledge of the subject but also its social, political, economic, emotional and ethical dimensions. He designs his lectures in such a manner that interweaves all the varied dimensions of a particular topic, such as GM and presents it in a holistic way to the students. Such kind of approach helps in fostering better interlinkages between the content and context of the learners and letting them own as well as apply their knowledge. He uses emotional metaphors while raising arguments related to

Table 9.4 Teacher's knowledge components constituting PCK

S.No.	Teacher's knowledge component (Code)	Evidence (in classroom interaction, discussion, interview, questionnaires, lesson plans, reflective fastwrites)	Frequency
1.	Subject matter Knowledge (SMK)	"Right …here they have taken a gene specifically the cry gene from a bacterium which produces a toxic protein and is toxic for insects especially the Boll worm that infects the cotton crop. The insect as soon as it infects the Bt-cotton immediately gets knocked off as the toxin is surface-specific."	42
2.	Teacher's Beliefs (TB)	"I believe a person becomes a person only after birth or during the latter part of gestation."	13
3.	Knowledge of Transactional Strategies (KTS)	After watching the documentary, a classroom discussion was being initiated by the professor so as to debate on the ethical consideration vis-à-vis animal transgenic technology	16
4.	Knowledge of the Curriculum (KCur)	"Ethics is something…… not that important and that's the reason these ethical issues have been removed from the syllabus. I personally feel that there should be one complete unit on Bioethics, as it is very necessary in the present day context. We need to sensitise not just our students but also our teachers so as to understand the inter-relationships that we share with the animal world."	5
5.	Knowledge of Students' Understanding (KSU)	"Some of them were very keenly interested in some of these ethical aspects. Also, when a teacher asks them that in a particular situation how would you react? They are ready to come out with their ideas. So, overall I think that there is interest in these matters"	1
6.	Knowledge of Context (KCon)	"I think it is more natural for all human beings to get more money, more benefits, etc. maybe that is what is driving all such sort of research in biotechnology."	14
7.	Knowledge of Assessment (KA)	"Go through the news items, books, and research articles for getting information about the spectrum of animals that have been modified and make a list of all such transgenic animals along with the purposes for which they have been modified. Are these animals modified to derive some	1

(continued)

Table 9.4 (continued)

S.No.	Teacher's knowledge component (Code)	Evidence (in classroom interaction, discussion, interview, questionnaires, lesson plans, reflective fastwrites)	Frequency
		benefit for humans or just about exciting the aesthetics, such as is the case with animals glowing in the dark? Think about this, "Is it morally right for modifying such organisms?"	
8.	Knowledge of ethical theories/Role of ethics (in general and in Science) (KET)	"All such ethical issues and the ethics involved therein needs to be discussed as we are living in a society, and we are bound by certain rules and regulations. Most of these laws are self-imposed, as human beings make these laws, and nobody is born to follow these laws."	5
9.	Orientation toward Science Teaching (OST)	"The eventual goal of science education is to produce individuals capable of understanding and evaluating information that is, or claims to be, scientific in nature and of making decisions that incorporate that information appropriately and, furthermore, to produce a sufficient number and diversity of skilled and motivated future scientists, engineers, and other science-based professionals"	3
10.	Knowledge of Ethical Argumentation (raising arguments/ initiating argumentation/ resolving) (KArg)	"The critical point over here is that this company has gotten a monopoly over this particular seed type and has got a patent for it. And if you have to use it, then you have to buy the seed from Monsanto."	45
11.	Knowledge about Nature of Science (KNOS)		22
		PCK_{TOTAL}	**167**

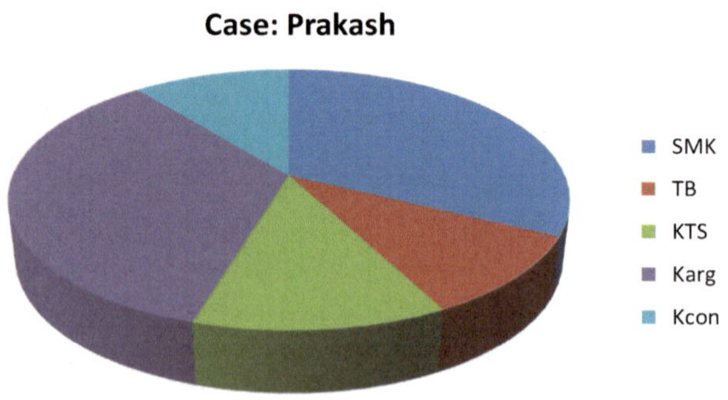

Fig. 9.1 Graphical representation of components of PCK

various technologies in order to develop ethical understanding among the learners. At the interpersonal level also he shares a good rapport with his students, which is observable, felt and reflected in the way he interacts and gels with his students. In a reflective note, he has reflected upon this facet of his personality,

> Being a rather sociable sort of a guy, I found that I could understand students' mentality... could get under their skin so to say....and was able to establish an easy, informal relationship with them. I quickly found that students were comfortable with a teacher-friend, one they accepted as their own. (Reflective Note)

The teacher feels that knowing one's students is very important for delivering an effective lesson which is one of the most important attributes of PCK.

> Within the context of burgeoning numbers in the classroom, one has to make an effort to know one's students....who is a quick learner and who is not. This is important because the approach to the two will be different...and you are not a good teacher unless you can reach both. (Reflective Note)

Thus, the pedagogy should be well adjusted to the diverse needs of the learners and cannot be generalized for every learner in the class. Hence, the teacher is well acquainted with the different contextual factors that put an impact on teaching and makes an effort to maximize his teaching potential.

9.12 Conclusion

The present teacher holds contextual teaching and learning as near to his approach of teaching ethical issues in Biological Sciences. According to him, Science cannot be taught effectively without contextualizing its theory, principles and method. The term 'context' originated linguistically from the term 'Contextere' which means to weave something together or a relationship (Podschuweit & Bernholt, 2018). Duranti and Goodwin (1992) defined Context as, "a relationship between two orders of phenomenon (a focal event and a field of action) that mutually inform each other to comprise a larger whole." Context-based approach in Sciences is similar to the STS approach and is found to be more effective as compared to traditional science teaching learning practices (Bennett, Lubben, & Hogarth, 2007). It has been proven that a students' conceptual growth of a scientific concept is greater when she/he can apply in different contexts (Podschuweit & Bernholt, 2018). One of the characteristic feature of context-based approach is 'pupil-centred learning', participatory learning or called as active learning. Examples of such learning activities include small group discussions, group and individual problem-solving tasks, investigations and role-plays (Bennett, 2003). Thus, the present teacher uses many such exercises, experiences along with emotional metaphors to make learning contextual for his learners. Although difficult to practice, context-based approach to teaching has greater pay-off as compared to the traditional science teaching, as the students gain an authentic understanding of a scientific concept and are able to simultaneously apply their learning in different contexts. For the teaching of difficult science concepts, contextualized approach could be promoted and students should be actively engaged in the activities that help in building knowledge. Teachers can specifically design contextual-learning tasks for their students that aid in better understanding of the Science concept or phenomenon.

References

Bennett, J. (2003). *Teaching and learning science*. New York: Continuum.

Bennett, J., Lubben, F., & Hogarth, S. (2007). Bringing science to life: A synthesis of the research evidence on the effects of context-based and STS approaches to science teaching. *Science Education, 91,* 347–370.

Duranti, A., & Goodwin, C. (1992). *Rethinking context: Language as an interactive phenomenon*. Cambridge: Cambridge University Press.

Podschuweit, S., & Bernholt, S. (2018). Composition-effects of context-based learning opportunities on students' understanding of energy. *Research in Science Education, 48,* 717–752.

Saban, A., Kocbeker, B. N., & Saban, A. (2007). Prospective teachers' conceptions of teaching and learning revealed through metaphor analysis. *Learning and Instruction, 17*(2), 123–139. https://doi.org/10.1016/j.learninstruc.2007.01.003

Saban, A. (2010). Prospective teachers' metaphorical conceptualizations of learner. *Teaching and Teacher Education, 26* (2), 290–305.

Toulmin, S. (1958). *The uses of argument*. New York: Cambridge University Press.

Toulmin, S. (2003). *The uses of argument* (Updated ed.). U.K: Cambridge University Press.

Part III
Ethics in Science-Tools for Learning

Chapter 10
Ethical Understanding Among Students About Bioethical Issues

Abstract Students have varied perceptions about ethical issues related to Biotechnology which was evident from their responses in the classrooms at both high school and undergraduate levels. One of the factors contributing to the development of students' ethical understanding is the quality of exposure that they are given in the classroom vis-a-vis ethical issues. The classroom discourse initiated by the teacher around these issues also helps in triggering thought process and critical understanding about these issues. The present chapter highlights the different aspects of students' thinking, meaning-making and ethical understanding with respect to STSE (science, technology, society and environment) issues which is often evident in their arguments. Sometimes, it was found that students abide by their teacher's positions and stances on ethical issues, while in others, there were also some conflicts recorded in their responses. This will give a firsthand understanding to the teacher as to how students think about ethical issues in Biotechnology and can thus help them in better planning and transaction of issues. Development of ethical reasoning is indispensable for sound decision-making and attaining a high level of scientific literacy on these issues, which was reflected in some students' responses.

Keywords Ethical issues · Biotechnology · Ethical understanding · Argument · Ethical reasoning

10.1 Introduction

High school and Undergrad Biology courses now include topics related to Biotechnology, such as stem cell technology, genetic engineering, human genomic project, cloning, amniocentesis, in vitro fertilization (IVF), organ donation, production of pharmaceutical drugs in animals, etc. (Cavanagh, Hood, & Wilkinson, 2005). The understanding with respect to these topics is not just based on content/concept clarity rather has an ethical dimension to it. These topics are usually covered under a separate branch of Biotechnology referred to as 'Bioethics' since it

A. Saxena, *Ethics in Science*, https://doi.org/10.1007/978-981-32-9009-9_10

involves the human element and technologies that impact and are impacted by social, political, religious, cultural and ethical contexts at a given time in any society. In different countries, the teachers do discuss these Bioethical issues in the classroom and generate a healthy conversation and debates (Macer, Bezar, Harman, Kamada, & Macer, 1997), but how much effective it is in generating an authentic ethical understanding amongst students is still a big question. Therefore, it becomes imperative to understand the students' preconceived notions and beliefs about these Biotechnologies so as to detect any misconceptions or flawed notions, as our aim is towards building authentic scientific literacy amongst each and every individual. It has been agreed upon by the researchers that students' opinions about Biotechnology are deeply affected by the way in which the concepts and content are presented to them as well as exposure to different sets of discussions and argumentations on socio-scientific issues (Macer et al., 1997; Macer, Azariah, & Srinives, 2000; Bottcher & Meisert, 2013; Wang, Hong, Liu, & Lin, 2018).

In order to gauge the students' ethical understanding of these Bioethical issues, a qualitative analysis was carried out. Students' responses were recorded in the classroom scenarios, as well as after class scenarios in FGDs (Focus Group Discussions), at both high school and undergraduate levels. The responses from around 100 students at the high school level and 200 students at the undergraduate level were subjected to content analysis, and the major themes were drawn out. The responses at both high school and undergraduate levels are presented separately so as to identify the key points of similarity and differences.

10.2 Teachers Possess a Generic Understanding of Students' Conceptions Related to Ethical Issues in Biological Sciences

Teachers at both levels hold a general understanding of their students' conceptions and misconceptions related to ethical issues in biological sciences. Teachers at class-XII level think that students are informed about new technologies; however, they do not have a critical view of the same. They attribute this to their 'immaturity' and inability to formulate a balanced perspective. Some teachers also believe that the realm of ethics is beyond the comprehension level of students and they are unaware of its relevance in the decision-making process. Teachers at undergraduate level hold an encouraging notion about their students' understanding and think that students these days are very much aware of ethical issues. Teachers feel that science students are able to weigh the pros and cons of a particular technology and understand the issues better than others. But still, according to them, the students are not able to take a firm stand when it comes to decision-making vis-à-vis ethical issues. The teachers at UG level attribute this to the level of abstraction attached to these issues. Further, a fewer number of practical activities and experiments limit students' understanding of these issues.

Thus, teachers at both levels provided a general understanding of their students' views on ethical issues. However, their inability to provide specific arguments that students use points to their lack of awareness of their students' conceptions and ethical stances. Some studies (Clermont, Borko, & Krajcik, 1994; Van Driel, Verloop, & De Vos, 1998) have shown that teachers' knowledge of students' conceptions and difficulties do have an impact over their PCK and enhances their PCK. But some studies (Usak, 2009) have revealed that prospective teachers are not much aware of students' difficulties and misconceptions. This negatively impacts the development of ethical understanding amongst students with respect to issues related to technology. Research conducted by Jones, McKim, and Reiss (2010) shows that teachers may find it intimidating to take up ethical issues in the science classrooms as this could result in the need to rebuild their own concepts of nature of science and technology. Sometimes teachers were not able to specify problems faced by their students in understanding ethical issues and were not aware of their students' preconceived notions and level of ethical understanding. A number of studies (De Jong & Van Driel, 2001 cited in Friedrichsen, Abell, Pareja, Brown, Lankford, and Volkmann, 2009) have also highlighted the fact that pre-service science teachers did not show concern for students' learning prior to planning. This clearly reinstates the importance of a guided framework for negotiating ethical issues in the classroom and developing ethical reasoning amongst the students. This requires effort both at pre-service and at in-service levels for integrating ethical concerns in the curriculum.

10.3 A Roadmap for Ethical Reasoning

Unlike Science and scientific research, the matters related to ethical issues in Biotechnology cannot be resolved by pure objective means or a 'Scientific Method'. Since 'Ethics' is a branch of Philosophy, the methods of comprehension and processes of resolution of a problem are different. This branch of Philosophy called 'Axiology' includes 'Ethics' as a subset of it and pertains to a set of moral principles that govern one's behaviour or conducting of an activity, differentiating between 'good' and 'evil' along with the reasoning behind the same. Three major areas of study can be recognized[1]:

- *Meta-ethics*: It is the study of the kind of things that exist in the universe. Some things in this universe are made up of physical stuff, such as rocks, body and matter, as well as non-physical elements like thoughts, spirits and gods. This corresponds to the theoretical meaning and reference of moral propositions and how their truth value can be determined. It seeks to understand the nature of ethical principles, such as what is goodness? and how can we tell what is good or bad? Meta-ethics is the least precisely defined area of moral philosophy.

[1]Taken from, Internet Encyclopedia of Philosophy (https://www.iep.utm.edu/ethics/).

- *Normative Ethics*: This corresponds to a practical means for determining a moral course of action. It involves arriving at the moral standards that regulate right and wrong conduct. The various ethical theories under this field of ethics include the following:
  - Virtue ethical theory: This places less importance on learning the rules, but stresses upon the development of good character, and acquiring good habits, such as kindness, benevolence, sincerity, temperance and honesty.
  - Consequentialism (Utilitarianism): This gives more importance to the consequences of action which should be more favourable than unfavourable. Here, the end result of an action is usually the sole determiner of its morality.
  - Deontology: This refers to abiding by the duties and obligations that one has irrespective of the consequences that it may lead to. Another duty-based approach is the 'rights' perspective, such as human rights. The rights and duties are actually related to each other, such that the rights of one person imply the duty of another person.
- *Applied Ethics*: This refers to what a person is obligated to do in a particular situation or domain of action. It involves the analysis of specific, controversial ethico-moral issues, such as abortion, euthanasia and animal rights. It can be applied to real-world actions and moral considerations in the area of private and public life and professions, such as 'Bioethics Community' that comes under 'Applied Ethics'. Therefore, decision-making with respect to application of any Biotechnology will also come in the purview of applied ethics.

In the present book, since the focus is on the ethical issues related to Biotechnology, a recourse and reference to Applied Ethics will be more appropriate. However, normative ethics also forms an important part when it comes to individual decision-making and behaviour modification. Besides ethical issues related to Biotechnology, decision-making and course of our action in life-related circumstances, one often needs to rely on these ethical principles. Researchers have struggled with the idea of any existential or spiritual element that guides the people through challenging life dilemmas and some researchers have even talked about the role of moral intelligence in both children and adults (Gardner, 1999; Coles, 1998). Some people even believe that 'ethical reasoning' is irrational (Rogerson, Gottlieb, Handelsman, Knapp, & Younggren, 2011) on account of the fact that they do not abide by the Scientific Method. However, some researchers have developed a model of ethical reasoning wherein they argue that ethical behaviour or resolution of any ethical dilemma can only take place when an individual follows certain steps of an ethical reasoning process. The following steps have been proposed (Sternberg, 2012):

- Recognize that there is an event to which to react;
- Define the event as having an ethical dimension;
- Decide that the ethical dimension is of sufficient significance to merit an ethics-guided response;

- Take responsibility for generating an ethical solution to the problem;
- Figure out what abstract ethical rule(s) might apply to the problem;
- Decide how these abstract ethical rules actually apply to the problem so as to suggest a concrete solution;
- Prepare for possible repercussions of having acted in what one considers an ethical manner;
- Act.

These steps can be taken as a road map or sort of a guideline for reaching a decision that is ethically sound and is beneficial or non-maleficent for the larger population. However, as we shall see in the students' responses on the ethical issues, they may not be following a set route to reach a decision, rather go by hearsay and lop-sided view. This often leads to erroneous decisions which could bear harmful consequences for them, as well as for society. Therefore, it is pertinent to provide them with a framework, so as to scaffold them in the process of developing ethical reasoning.

The present chapter analyses students' responses and arguments on ethical issues at high school and undergraduate levels based on the ethical framework (as discussed in Chap. 1, also refer to Appendix G). The responses of the students were collected in two ways, either during the classroom discourse (in the presence of teacher) or as part of focussed group discussion (FGD).

10.4 Students' Perspectives on Ethical Issues at High School Level

10.4.1 Students' of Arunima's Class

The students in Arunima's class were allowed to watch a video on "Monsanto and Indian farmer suicide" (https://www.youtube.com/watch?v=Av6dx9yNiCA). After watching the video, a focussed group discussion was organized and students were asked to give their feedback and comments on the GM technology. Some of the arguments raised by them clearly reflect their understanding about the whole issue of GM crops.

> More such technologies should be launched so that farmers can earn enough money for a better living. GM crops are good and beneficial for both agriculture as well as farmers. (Pro-technology & utilitarian)

> I personally believe that there is no harm in growing Bt-Cotton. It is just a hybrid seed in which all the parental genes are the same but also a pest-resistant gene is introduced in it, and if they are well irrigated, surely they can help and benefit the farmers. (Pro-technology)

> GM crops are good because it will help in the production of vegetables which have more nutritional value. (Utilitarian)

> Technology like GM crops can help in long run in preventing environmental pollution by pesticides. However, for the proper utilization of such techniques other factors hindering the benefits caused by them should be checked. (Somewhat balanced viewpoint)

> GM crops are beneficial but it needs constant monitoring and help from the Government and scientific help to the peasants. (Somewhat balanced viewpoint)

> GM crops are helpful as it is resistant to insects like of order Lepidoptera. It also increases the yield of cotton. Yes, I agree with the production of GM crops. (Pro-technology)

Thus, most of the students tend to see the beneficial side of the technology of genetic modification and give different arguments in its support. Some of them seem to be completely unaware of the negative side that such a technology can have. They highlighted the role of government authorities and scientists in monitoring the effective implementation of the technology. But they are unable to view the issue from an ethical perspective, the various harmful effects that the technology can have on humans, environment, ecosystem balance, species diversity, the emergence of superweeds, etc.

10.4.2 Students' of Saira's Class

The students in Saira's class seemed to be well read and aware of the whole issue of GM, which was evident from the quality of their responses, including their personal beliefs and stand taken on the issue. The major themes that emerged from students' responses have been presented below (refer to Table 10.1).

As is evident from the responses, most of the students were able to provide well-reasoned arguments on the whole issue of GM technology. This was especially because they are aware of the benefits and costs incurred by the Bt-technology and are able to analyse the conflicting arguments posed by different groups of people. For example,

> Bt-technology is bad because it produces severe effect on farmer's income, can promote resistant pests and affect the ecology of the area.

In one of the student responses, the stand taken by the student clearly reflects sound knowledge of the whole issue.

> Biotechnology if used in a wrong way can be harmful for the people. Irrigation facilities are not provided to the farmers and they are in a big loss. Technologies should be made by keeping poor farmers and their conditions in mind. They should be told about the consequence and what all things the crops require.

A focussed group discussion on the issue of IVF resulted in following students' responses:

—How many of you are in favour of this technique of In vitro fertilization?
[Few students raised their hands, but there appeared a slight hesitation on the students' part.]
—(question posed to those who had raised their hands) Why do you think so?

Table 10.1 Students' responses over the issue of GM crops

GM crops are beneficial	GM crops need proper and stricter regulation	Public knowledge and understanding of GM crops is important	Bt-technology is unsafe	Bt-technology is selective	Vested interest in promotion of GM crops
"Bt-technology is highly useful and boon to mankind but conditions apply." "Bt is good and can be used to increase theproduction." "Bt- technology, if used with proper knowledge, can make farmers' condition better." "GM crops are better as they are disease and pest-resistant."	"GM crops should be strictly field tested." "Proper study, experimentation and investigation must be carried out so that the masses are not affected." "Bt-cotton should not be abolished completely but tried, tested and released for betterment."	"Farmers need to understand and weigh the pros and cons of Bt-technology and only then come to a conclusion."	"Crops should be naturally grown as any kind of manipulation may cause harm."	"The crops should be engineered for every climate condition so as to give the best output." "Bt-crops are not beneficial for the Indian farmers as they do not get required yield." "GM crops might give good results only in certain environmental conditions."	"Monsanto is behaving like a business company playing with the farmers' lives."

St 1: Because it provides a ray of hope to the infertile couples.
St 2: It is a boon for those cannot conceive and grants them happiness.

Thus, from the classroom discussion and focussed group discussion with the students, it can be said that students have a uni-dimensional view with regard to the technology. Therefore, they are unable to generate any argument even when a number of ethical questions were posed to them. This can be taken as a misconception on the part of students that should have been dealt by the teacher, substantiating with some evidence or examples. She could not carry out a detailed analysis of students' notions and ideas about technology. One of the major ethical issues associated with the technology of IVF relates to embryo wastage during the procedure of IVF and then after the cryo-preservation of excess embryos. Embryos have a human potential and can give rise to a complete human life and the procedure of IVF subjects to the killing of many human lives which is not acceptable on ethical and moral grounds. But the teacher did not bring this issue to the fore which reflects her limited understanding vis-à-vis the technology and its implications on human society. This is a clear reflection on the pedagogy and PCK component of the teacher wherein her subject matter knowledge (SMK) with respect to ethical issues is very limited and is rather simplistic.

In another topic pertaining to Human Genomic Project, certain ethical issues were brought to the fore (refer to Box 10.1), such as the interference posed by HGP in the process of evolution and loss of differentiation and diversity amongst organisms if genetic selection is being carried out so as to give rise to perfect individuals.

Box 10.1 Classroom Discourse on Human Genomic Project (HGP)

Teacher:	What do you understand by HGP?
Student 1:	To find the genetic code and arrangement of base pairs.
Student 2:	Many countries collaborated in this project.
Student 3:	The cost incurred was around 3 USD per base pair.
Teacher:	What was the need for such an expensive project?
Student 4:	Genes can cure diseases.
Student 5:	To find the number of functional genes.
Student 6:	To get "better species".
Teacher:	Do you think that we should go for "better species"? Who decides that which species are 'better'? (KArg)
Student 7:	Yes, I think we should go for it if we have better chances of survival or in any way beneficial for the human race. (Consequentialist)
Teacher:	Then what will happen to nature's own process of evolution? (Consequentialist) (KArg)
Student 8:	Through this process evolution can occur at a faster pace.
Teacher:	What are the advantages and disadvantages of HGP?
Student 9:	There are no disadvantages.

Student 10:	People with destructive thinking can create certain destructive organisms that can harm the society as well as the environment.
Student 11	This can lead to the perpetuation of some injectible mutation as well as risk an epidemic break out. (Consequentialist)
Student 12:	Certain cloning issues will arise.
Student 13:	This technique can be misused by terrorists. (Consequentialist)
Teacher:	The basis for evolution is differentiation amongst organisms, and if HGP allows us to select our genes or genes of the offsprings, for instance getting them the perfect genes for intelligence, physical beauty, strength and athletic prowess then there would be no competition amongst species thus slowing down the process of evolution. (Consequentialist) (KArg + SMK)
Student 14:	It depends upon how we use the technology.
Teacher:	Who is going to decide that?

However, a point that was put forward here was the behaviour of a particular person which is not a result of just the intrinsic or genetic factors but how the extraneous factors also play a role in determining the characteristics of an individual. Another issue of ethical concern discussed by the teacher was the informed consent of the individual whose genome is taken for deciphering the base pair sequence. Besides these, the focus was also laid onto the conceptual and technical details of HGP, such as coding and non-coding part of DNA, significance of satellite DNA in deciphering the human genome, role of Junk DNA and introduction of Bioinformatics as a new branch of study where software are being developed for finding the sequence of genome.

The class scored high on the ethical pointer scale but lagged behind on account of lower conceptual clarity.

10.5 Students' Perspectives on Ethical Issues at Undergraduate Level

10.5.1 Students of Meena's Class

A focussed group discussion with students of Meena's class was carried out (refer to Box 10.2) after they studied the topic of Genetically Modified Crops. It helped in gauging students' understanding of GM crops.

Box 10.2 Focussed Group Discussion on Genetically Modified (GM) Crops

—	What is your general take on GM crops and GM technology per se?
Student 1:	Although we have studied about GM crops, but in the long run we do not know what will be their impact on human health.
Student 2:	But these crops are certainly known to increase the production level.
Student 3:	We have inserted a foreign gene into the plant which can lead to any kind of health hazard or an adverse impact on the environment.
Student 4:	Since Bt-cotton was not edible but Bt-Brinjal is edible, due to which a lot of controversies have been attached to it.
—	Looking at the recent researches and recent reports, what do you think is the future of Bt-Cotton in India? Has it been a success or a failure?
Student 4,5,6,7:	It has been a failure.
—	Why do you think so?
Student 5:	The Bt-gene could escape in the environment.
Student 6:	The soil fertility also got decreased.
Student 7:	Increase in farmers' suicidal tendency because of loss of crop.
—	Now, I will narrate a case story, please listen to it carefully, as I want your comments and responses on the same. In a nearby general store you can find different varieties of Brinjal, such as normally grown variety wherein pesticides and fertilizers are used to increase the yield and size of the crop, another variety of Brinjal that is grown without using any chemical pesticides or fertilizers (or organically grown), besides there is still another variety of brinjal that is grown without using pesticides but contains a pest-resistant toxin gene (or is genetically modified). Which brinjal variety would you prefer to buy as a consumer and why?

Student 5,6,7,8, 9,10 (collectively):	Organic variety of Brinjal.
—	Why do you think so?
Student 7:	They will be devoid of any pesticide or fertilizer effect and hence healthier to consume.
Student 8:	It will have least influence on the environment.
Student 10:	There is always this kind of fear with the genetically modified variety, wherein even one in thousand brinjals, if carries a mutant transgene or a transgene that was not accurately synthesized, if consumed could lead to disastrous consequences.
—	Can organic farming support the entire human population, even if it is limited to the human population inhabiting the Indian subcontinent?
Student 6:	(no valid argument from students' side)
—	What should be done then? Are there any takers for Bt-crops?
Student 7:	(no response, appears as if they are either in a state of a dilemma or have gone anti-technology)
—	What could be a solution for all this? What can be done to resolve this issue of GM?
Student 2:	Some regulatory bodies can be set up that monitor the quality and field trials of GM crops.
Student 4:	The suppliers of these GM seeds or crops should ensure the farmer or the consumer the safety of the crop.
Student 5:	The crops should not be produced in bulk but in a limited quantity so as to monitor their impact on environment and human health.
Student 11:	What about GM- rice or golden rice that are said to be a precursor of beta-carotene and are so much useful for feeding the malnutritioned population?

The responses of the students indicate that although they are able to think about the different consequences of GM technology, they are not in a position to reach a consensus or balanced opinion. The arguments which they present are sometimes well reasoned and supported with evidence and theoretical backing, but at other times, they are also pure claims. Students were able to understand and reflect on these issues and offer solutions also. The students share diverse opinions about GM crops, and they also share their fears and dilemmas with respect to the impact of such a technology on human health, soil fertility and environment. They are not ready to buy a GM food crop as a consumer and prefer the organic crop over it. But there is an ethical dilemma that still persists such that a few of them were able to enumerate certain benefits of GM crops in the form of 'Golden Rice'.

10.5.2 Students of Kunal's Class

The topic chosen for collecting and understanding students' beliefs and opinions happened to be Transgenic Organisms and the Biotechnological procedures in general. A focussed group discussion on each of these issues was conducted so as to generate students' opinions and viewpoints. The following questions were asked to initiate the discussion (refer to Table 10.2):

- What is the need for the production of GM crops in India?
- Are GM crops safe to grow or consume?
- Is Biotechnology always used for the welfare of human society?
- Why do we create transgenic organisms? Where is the need?
- What if these transgenic organisms escape the lab into the environment?
- How can we ensure appropriate safety standards and regulatory structures?

The above discussion on Genetically Modified organisms depicts the multiple dimensions of the issue of GM, which is not just a science or technology issue rather more of an ethico-moral issue as well. The students' arguments reveal that they are sensitive and aware of the harm that such a technology could offer, such as brutality against organisms, impact on non-target species and imbalance in the ecosystem. In order to gauge the decision-making abilities of the students with respect to Biotechnology and its applications, the following discussion trigger was provided to them for further discussion and deliberation.

Issue 1 Trigger: Case story on GM crops
In a nearby general store you can find different varieties of Brinjal, such as normally grown variety wherein pesticides and fertilizers are used to increase the yield and size of the crop, another variety of Brinjal that is grown without using any chemical pesticides or fertilizers (or organically grown), besides there is still another variety of Brinjal that is grown without using pesticides but contains a pest-resistant toxin gene (or is genetically modified).

Table 10.2 Students' arguments on ethical issues (Topic: Genetically Modified Organisms)

S.No.	Arguments raised by students	Nature of the argument	Related ethical theory
1.	Certain biotechnological techniques have been quite useful for human welfare, such as in the sphere of medicines and pharmaceuticals	Informed argument with one justification	Utilitarian
2.	There are many limitations	Naive argument	
3.	Till now, we have been studying what all is happening inside the organisms, but with biotechnology, we can even modify the organisms to better its chances of survival	Informed argument with one justification	Utilitarian
4.	That is because there is a very thin line of difference between use and exploitation, and we should make use of any technology within certain boundaries	Normative argument	Deontology
5.	It is we humans who are using it, and we have to see that to what limits can we use a technology so as to cause the least harm to the people. It can be done only by observations, as these things cannot be solved theoretically	Intermediary argument	Consequentialist
6.	We usually study the success stories of a particular technology, we should also look into what could be the possible drawbacks of a particular technology	Intermediary argument	Consequentialist
7.	For human welfare	Subjective argument	Utilitarian
8.	To increase the productivity	Informed Subjective argument	Utilitarian
9.	Increasing the milk production of cattle	Informed Subjective argument	Utilitarian
10.	Along with increase in productivity, our economy will also grow	Subjective argument	Utilitarian
11.	First of all, proper sterile conditions have to be maintained, the entry to the lab will be restricted. The unsuccessful attempts wherein a desired organism is not produced will then be autoclaved	Expert view	
12.	One thing that I would like to add is that creating and destroying so many organisms can give rise to a brutality in us, which is not good	Subjective argument	Ethics of care for living beings
13.	Sometimes, the non-target species also get affected which could lead to imbalance in the ecosystem	Informed Subjective argument	Consequentialist

Questions for Discussion

- Which Brinjal variety would you prefer to buy as a consumer and why?
- Do you think that Indian farmers are well equipped to grow and cultivate GM crops or have the requisite expertise and knowledge in this area?
- What are the benefits of creating genetically modified organisms?
- What could be the consequences if these GMOs escape in the environment?

The following arguments were generated by the students based on the above situation (refer to Table 10.3).

Thus, the arguments generated by the students on general Biotechnological procedures, as well as 'Transgenic organisms (GMOs and GM crops)', usually fall under utilitarian perspective, as they see the beneficial side of technology, which is for the welfare of humankind. Sometimes the arguments are based on sound justification, while most of the times, they are naïve and of an intermediary kind. This means that their arguments are usually based on subjective opinions or hearsay without any empirical evidence for it. Students were able to identify and foresee both the positive and negative impacts that a technology can bring about and the regulatory provisions that need to be in place for a sustainable and healthy environment, such as

> Certain biotechnological techniques have been quite useful for human welfare such as in the sphere of medicines and pharmaceuticals. *(Pro-technology argument)*
>
> It is we humans who are using it, and we have to see that to what limits can we use a technology so as to cause the least harm to the people. It can be done only by observations, as these things cannot be solved theoretically. *(Technology regulation)*
>
> One thing that I would like to add is that creating and destroying so many organisms can give rise to a brutality in us, which is not good. *(Anti-technology)*
>
> Sometimes, the non-target species also get affected which could lead to imbalance in the ecosystem. *(Anti-technology)*
>
> The foreign gene may get integrated with the human genome and play havoc with our systems, or cause certain health hazards. *(Anti-technology)*

From the above account, it is evident that at least some of the students are aware of the ethical concerns associated with GM technology but are unable to put forth reasoned arguments and a balanced perspective with respect to technology. This shows that not enough discussion on ethical issues goes on in the class in order to substantiate and support students' claims, as well as question their naïve ideas.

Table 10.3 Students' arguments on GM crops

S.No.	Arguments raised by students	Nature of the argument	Related ethical theory
1.	I will go with the second one	Subjective argument	
2.	If anybody's pocket allows, then definitely he/she will go for the second one	Informed argument with one justification	
3.	Certainly first variety	Subjective argument	
4.	The foreign gene may get integrated with the human genome and play havoc with our systems or cause certain health hazards	Informed argument with one justification	Consequentialist
5.	These researches grow up to bring out some positive effects from these crops. And 20 years or 30 years down the line, the researchers may be able to convince the masses by telling them about the benefits that these crops have to offer, because, with the ever-increasing size of population, the problem of food security will surely come and for which only GM can provide a viable solution	Informed subjective argument with more than one justification	Utilitarian
6.	I think that awareness of these GM crops needs to be generated to convince the people	Subjective argument	
7.	Since with organic farming only a small and selected population can be supported, we are only left with nothing but GM	Intermediary argument	
8.	Genetic engineering is the best possible technique available to us at this moment and all hopes are set on it	Naïve argument	
9.	Any day, as the research is more important, and without it, the cures cannot be discovered	Informed argument with one justification	Utilitarian
10.	If you are saving ten lives of human beings by sacrificing two mice, then I think it is justifiable	Naïve argument	Human versus Animal life-Deontologic ethic

10.5.3 Students of Prakash's Class

The topics chosen for gauging students' beliefs and ethical understanding were GM crops and Contraception. A focussed group discussion on each of these topics was initiated so as to generate students' opinions, concerns and viewpoints and allow them to argue with each other in a healthy debate kind of setup. The following arguments from students reflect their understanding of the various ethical issues attached to these technologies (refer to Table 10.4).

Table 10.4 Students' general arguments on GM crops

S.No.	Arguments raised by students	Nature of the argument	Related ethical theory
1.	One needs at least to be informed about what exactly GM crops are, as only then we would be able to make an informed choice regarding which crop to buy	An intermediary view	Principle of Informed Consent
2.	Increasing the yield usually in turn means increasing the economics or increasing the financial benefits; it has not been happening at least with the Bt-thing because the Andhra Pradesh farmers did commit suicide after being entrapped by the foreign multinational companies which are just looking for their own profit and all that terminator seed technology is an offshoot of that. So, it is the Govt. that should look into the matter and these 'ethics' should be taken care of	Counterargument with more than one justification (an informed view)	Principle of Justice/ Autonomy
3.	Right now, we are going through a hit-and-trial phase in lieu of increasing the crop yield or for the economic growth of our country. And now also, so many people are starving and dying of hunger, so where are those crops and the claims that are being made for supporting such crops?	Counterargument with one justification (informed view)	Utilitarian
4.	I think here the farmers face the major problem, and they need to be trained in this regard as to how these GM crops should be grown. And here, only the Govt. can intervene	An Informed subjective view	
5.	The companies only project the 'good' aspect of this technology to sell their seeds and never tell the farmer about the 'terminator seed technology' which has been inserted into those seeds	Counterargument with one justification (informed view)	Consequentialist
6.	Am not sure whether the companies even have patents for such seeds, and companies, such as Monsanto, do have patents and have turned into major suppliers of such seeds and are trying to dominate the seed market, thus confiscating the free will of the farmers. And not everyone can afford such a technology as GM which comes with a lot of other products	Counterargument with more than one justifications (informed subjective view)	Principle of Autonomy

10.5.3.1 Issue 1: GM Crops (FGD)

From the above analysis of students' arguments, it can be understood that students are well informed about the ethical issues and are able to weigh them under different theoretical perspectives. The students are able to view the issues from multiple perspectives including, social, political, religious, economic, scientific and ethical perspectives. They are able to situate a particular issue in the given context and exhibit ethical reasoning on the same. In order to diagnose students' understanding and ethical decision-making skills, they were allowed to reflect on some case studies which were given in the form of 'triggers' (refer to trigger on GM crops mentioned above).

From the above arguments raised by the students (refer to Table 10.5), it can be said that they are well informed about the whole issue of Genetically Modified

Table 10.5 Students' arguments on GM crops

S.No.	Arguments raised by students	Nature of the argument	Related ethical theory
1.	The problem with such a kind of thing is that the most preferred option would be the 'organic crop' but not everybody can afford them	Subjective argument	Utilitarian
2.	The common man will go with the third crop variety which has been genetically modified, because they will not know that it is being genetically modified as they are not that informed and go by the appearance of a crop which is appealing in the case of GM crops	Informed argument with one justification	Consumer rights—Deontologic ethics
3.	The common man will go for the third one, because they already know that pesticides cause many side effects	Informed Empirical argument	Consequentialist
4.	Anyways, there are chemicals being ingested, whether we go for the pesticide-laced crops or GM crops, and I do not know but I've just heard that these GM crops were also so hurriedly introduced into the market without proper field testing on a larger population, so we cannot predict	Informed Empirical argument	Consequentialist
5.	Also, just testing them on few cannot tell the impact of such crops on the whole human society. Because there is something called "Bioprospecting", so if that is not done correctly, then we cannot get the correct impact on the population	Informed argument with one justification	Consequentialist

(continued)

Table 10.5 (continued)

S.No.	Arguments raised by students	Nature of the argument	Related ethical theory
6.	Since we are from Sciences, we know about these crops much more than a common man knows, but it will not make any difference to him whether it is a GM crop or a normally grown plant variety	Intermediate argument	Principle of informed consent
7.	As science students also, we are not going to be millionaires or billionaires, so we cannot go for the second variety and will go with the third one	Intermediate argument	
8.	I think many people will go with the first one as they think that the pesticides or fertilizers can be washed away and boiled out, so to annihilate their harmful impact, but some will go for the third one as they look good and are not that expensive as compared to the 'organic crop'	An ambivalent argument with one justification each	

crops and are aware of the consequences that they may pose to the farmer, the soil or land, the economy and society. They are able to think about the issue from varied perspectives and not just science and technology viewpoint. They are also socio-ethically aware of the pros and cons that these technologies have to offer. Many of them are sceptical in their arguments which clearly depicts that they are not ready to blindly follow whatever the science and technology has to offer, rather are critical of not only the very idea and intentions behind a particular technology but also its impact on the human society and environment. They still adhere to the natural means of growing the crops which is the organic way and regard it as safe and eco-friendly. They have informed consumers and are able to make choices for themselves, but at the same time, they also exhibit a concern for the larger population who may not be so self-sufficient.

10.5.3.2 Issue 2: Contraception

Issue 2 Trigger: Narration of a case story on 'Contraception'
Ethiopians who claim to be Jews are welcome to migrate to Israel under the Law of Return, but they face discrimination and have not always integrated well into Israeli society. Births among Ethiopian women have dropped by 50% in the last decade, according to a report by the "Vacuum" investigative news programme on Israeli Educational Television. "This story reeks of racism, paternalism and arrogance. It's a story to be ashamed of," journalist Gal Gabai concluded. Ethiopian women told the journalists stories of

unsubtle coercion and misinformation. "They said, 'Come, there are vaccinations, gather everyone," one of them said. "We said this is not a new problem, but the government is finally facing up to the lack of informed consent on the part of a marginalised, poorly-educated minority. In 2008, Hedva Eyal, of the feminist group Isha L'Isha, wrote a report alleging that the medical profession had failed Ethiopian migrant women. We wouldn't receive it. They said, 'You won't move to Israel.'" Women said that they were told that it would be hard for them to work or find accommodation if they had large families.

"The paternalistic attitude towards women of Ethiopian origin and the state's concern over high rates of birth among poor and black populations drove Israeli official bodies, such as The Jewish Agency and the medical establishment, to act, allegedly for the benefit of women's health, but in fact according to the concepts and wishes of the establishment regarding the desirable way to conduct family life. As a result, and as this paper shows, women did not get crucial medical information and their right of choice regarding their bodies, families and lives was severely curtailed."

- What is the argument for forced abortions in this case?
- Does the argument relate to the pro-life or pro-choice camp? Or is it merely an interplay between state regulation and a paternalistic attitude towards the women of Ethiopian origin?
- Is it a kind of racist and genocidal threat?
- How important is it to seek the consent of the pregnant mother for inducing abortion?
- Does the pregnant woman have the right to decide if she is going to carry the baby to term or not?

In the above discussion (refer to Table 10.6), the students are able to build a strong argument against forced abortions and MTPs carried out en masse by the Israeli people on Ethiopian women and even supported by the Israeli Government. The students claim that such a heinous act is completely against the pro-life camp. They also argued that abortion is a medical procedure that is being incurred to a woman's body, so she has a complete right over her body and some other person or government cannot deny her this reproductive right. The students are able to understand the role of state and politics in curbing the population but argue that such a practice is against human rights.

Table 10.6 Students' arguments on contraception

S.No.	Arguments raised by students	nature of the argument	Related ethical theory
1.	First, if the contraceptives are being given to all the women irrespective of whether they are pregnant or not, and if a pregnant woman is given a contraceptive, then it is definitely according to pro-life camp a wrong act	Informed argument with one justification	Ethics of informed consent
2.	Irrespective of whether a woman is pregnant or not, one needs to have knowledge about what is happening to one's body and an informed consent is absolutely important in this case	Informed argument with one justification	Ethics of informed consent
3.	It is something related to the communal rights of the Jewish people, as they cannot deny rights to the Jews in their country. And maybe they do not like black people, so they just want to end their society to reduce it to zero and to prevent mixture	Informed argument with one justification	Principle of equality of rights—Deontologic ethic

10.6 Conclusion

It is well evident from the above discussions that students do carry certain beliefs and notions about the recent Biotechnology and ethical issues associated with it. These notions and beliefs may be influenced by the understanding that they develop while studying the particular Biotechnology. Some of the students seem to be better informed than others in terms of formulation of reasoned arguments rather than citing pure claims. They are scientific in their thought process and rely on evidence and data to support their arguments. However, different cohorts of students display differences in their arguments and stance taking that can be attributed to different factors, such as students' background and previous knowledge about the ethical issues, sources of their information other than textbooks, level of awareness, maturity of thinking and weighing the issues, teachers' efforts and knowledge about the issues. Some students have ambivalent thinking when it comes to reasoning about the ethical concerns associated with various Biotechnologies (Knabe, Stearns, & Glasser, 1994). The extent of divergent thinking was evident in students' arguments, and they were able to think beyond the textbooks when confronted with some case studies or life situations. It depends upon how much they are exposed to such dilemmatic situations, where students need to think critically and apply ethical reasoning. However, at times they were even influenced by certain media reports and flawed portrayals which need to be checked and taken note by the teachers. Many a time, the students' responses reflect a reliance on different ethical theories and principles, such as Virtue theory, Utilitarian theory, Deontological theory, principle of beneficence and non-maleficence, but at times, their personal values conflict with societal norms and ethics (Iyalomhe, 2009). Having a correct and informed understanding of Bioethical issues is important for attaining scientific literacy and practice responsible Science. From the analysis of students' arguments at different levels, high school and undergraduate levels, as well as with respect to different schools and colleges, it appears that the quality of arguments varies and depends on the amount of debate and deliberations that happen at the classroom level. The teacher's role is again pertinent over here, although ethics cannot be taught to anyone, a teacher can provide opportunities and avenues for the students to share and exercise their opinion by the way of direct and indirect instructions (Bottcher & Meisert, 2013). There is a need for scaffolding by the teacher in order to assist the students in developing their ethical reasoning abilities and decision-making (Griswold & Chowning, 2013). This gives the student freedom of thought to generate one's own argument by weighing the issue from different perspectives.

Thus, the role of pedagogy is crucial to learning and decision-making with respect to ethical issues in Biotechnology. The difference is not so much in the grade level of the students as it is on the competency and knowledge of the teacher. As is evident from varied students' responses at K-12 and undergraduate levels, where the students are equally curious and keen to learn about these issues, however, the teacher's role and her pedagogy make all the difference in the classroom. At K-12 level, the teacher's main concern is about syllabus completion and preparing the students for the end-term examination wherein the focus on these ethical issues gets somewhere diffused, whereas at undergraduate level, the teachers are able to discuss these issues at length due to greater subject matter knowledge about these issues. This could be a reflection on the present teacher education programmes (both pre-service and in-service) that are insufficient to foster an understanding of the ethical issues related to Biotechnology and their pedagogy.

References

Bottcher, F., & Meisert, A. (2013). Effects of direct and indirect instruction on fostering decision-making competence in socioscientific issues. *Research in Science Education, 43*(2), 479–506. https://doi.org/10.1007/s11165-011-9271-0.

Cavanagh, H., Hood, J., & Wilkinson, J. (2005). Riverina high school students' views of biotechnology. *Electronic Journal of Biotechnology, 8*(2), 121–127.

Clermont, C. P., Borko, H., & Krajcik, J. S. (1994). Comparative study of the pedagogical content knowledge of experienced and novice chemical demonstrators. *Journal of Research in Science Teaching, 31,* 419–441.

Coles, R. (1998). *The moral intelligence of children: How to raise a moral child.* New York, NY: Plume.

De Jong, O., & Van Driel, J. H. (2001). The development of prospective teachers' concerns about teaching chemistry topics at a macro-micro-symbolic interface. In H. Behrednt, H. Dahncke, R. Duit, W. Graber, M. Komorek, A. Kross, & P. Reiska (Eds.), *Research in science education: Past, present and future* (pp. 271–276). The Netherlands: Kluwer Academic.

Friedrichsen, P., Abell, S. K., Pareja, E. M., Brown, P. L., Lankford, D. M., & Volkmann, M. J. (2009). Does teaching experience matter? Examining biology teachers' prior knowledge for teaching in an alternative certification program. *Journal of Research in Science Teaching, 46* (4), 357–383.

Gardner, H. (1999). Are there additional intelligences? The case for naturalist, spiritual, and existential intelligences. In J. Kane (Ed.), *Education, information, and transformation* (pp. 111–131). Upper Saddle River, NJ: Prentice Hall.

Griswold, J. C., & Chowning, J. T. (2013, Spring). Strategies to support ethical reasoning in student argumentation. *Issues in Teacher Education, 22*(1), 63–75. Retrieved from https://files.eric.ed.gov/fulltext/EJ1013926.pdf.

Iyalomhe, G. B. (2009). Medical ethics and ethical dilemmas. *PubMed* (pp. 8–16). Retrieved from https://www.ncbi.nlm.nih.gov/pubmed/19485140.

Jones, A., McKim, A., & Reiss, M. (2010). Towards introducing ethical thinking in the classroom-Beyond Rhetoric. In A. Jones, A. McKim, & M. Reiss (Eds.), *Ethics in the science and technology classroom* (pp. 1–6). Netherlands: Sense Publishers.

Knabe, B., Stearns, J., & Glasser, M. (1994). Medical students' understanding of ethical issues in the ambulatory settings. *Family Medicine, 26*(7), 442–446.

Macer, D., Bezar, H., Harman, N., Kamada, H., & Macer, N. (1997). Attitudes to biotechnology in New Zealand and Japan in 1997, with international comparisons. *Eubios Journal of Asian and International Bioethics, 7,* 137–151.

Macer, D., Azariah, J., & Srinives, P. (2000). Attitudes to biotechnology in Asia. *International Journal of Biotechnology, 2*(4), 313–332.

Rogerson, M. D., Gottlieb, M. C., Handelsman, M. M., Knapp, S., & Younggren, J. (2011). Nonrational processes in ethical decision making. *American Psychologist, 66,* 614–623.

Sternberg, R. J. (2012). A model for ethical reasoning. *Review of General Psychology, 16*(4), 319–326.

Usak, M. (2009). Preservice science and technology teachers' pedagoical content knowledge on cell topics. *Educational Sciences: Theory and Practice, 9*(4), 2033–2046. Retrieved from http://files.eric.ed.gov/fulltext/EJ867376.pdf.

Van Driel, J. H., Verloop, N., & De Vos, W. (1998). Developing science teachers' pedagogical content knowledge. *Journal of Research in Science Teaching, 35,* 673–695.

Wang, H., Hong, Z., Liu, S., & Lin, H. (2018). The impact of socio-scientific issue discussions on student environmentalism. *EURASIA Journal of Mathematics, Science and Technology Education, 14*(12), 1–15. https://doi.org/10.29333/ejmste/95134.

Chapter 11
Assessment of Ethical Understanding—Tools, Techniques and Strategies

Abstract It is being said that 'ethics' cannot be taught; therefore, the question of assessing its understanding also stands redundant. But ethical understanding is often reflected through different activities and behaviours in the daily lives of the people. This chapter focuses upon the different ways in which a teacher can assess a student's ethical understanding with respect to the ethical issues in the most implicit way. It involves both formative and summative assessment criteria and can be individualized as well, depending on the need and level of the learners. The aim is not only to assess the students' ethical understanding but more so to develop it to higher levels of thought, application and reasoning. The chapter contains some tools and techniques, such as worksheets, learning log, group activities, projects and e-portfolios that could assist the teachers in planning their assessment strategies and also enable our young learners in their self-assessment.

Keywords Ethical understanding · Formative assessment · Summative assessment · Self-assessment

11.1 Introduction

Assessment is an important and indispensable element of teaching–learning cycle. It holds a significant place not just for accomplishing certain set standards and benchmarks for the students but is even more helpful and directional for the teachers in planning their teaching in a better way and also for the learners to implement better learning styles and grow in the process. Pedagogy and Assessment are intimately connected, rather than complementing each other. Tyler (1949) also emphasized more on the role of evaluation[1] in ascertaining the pedagogy and improving upon it. In order to understand whether the learners have actually attained the requisite levels of understanding as per the stage or grade level, assessment has to be done. The recent

[1]In the present chapter, the terms 'Assessment' and 'Evaluation' are used where 'assessment' refers to a classroom research that informs the process of teaching and learning. 'Evaluation' is solely carried out for the purpose of grading and reporting of student progress.

A. Saxena, *Ethics in Science*, https://doi.org/10.1007/978-981-32-9009-9_11

research literature on Assessment has testified the role of both formative and summative assessments for achieving the desired learning standards. Since the focus of the present chapter is on the assessment of ethical understanding, certain strategies and criteria that will be discussed here may not apply for other content areas. There has been a dearth of literature that addresses and explains students' mastery over ethical concepts and their organization of ethics knowledge (Jones, van Kessel, Swisher, Beckstead, & Edwards, 2014). Ethical understanding comes under the ambit of the affective domain, the assessment of which can only be done more through implicit and subtle ways as it deals with the inner belief systems, assumptions and at times emotions. There has been much deliberation and little consensus on the ways of assessing ethical reasoning amongst the students (Swisher, 2007). Different methods used by ethics educators include case or vignette analysis, objective examinations, surveys, focus groups and personal reflections (Jones et al., 2014). Assessment of ethical understanding will also entail the acquisition of decision-making capacities and certain virtues for the betterment of society and social responsibility.

11.2 Criteria for Ethical Understanding

Ethical understanding is mainly concerned with the development of critical thinking ability and reasoning as to what ought to be done and what ought not to be done. In order to assess the learners on their ethical understanding, first of all, there arises a need to know the standards or criteria that we are aiming at. Certain general criteria that can be followed for assessing ethical understanding amongst learners are provided as under[2]:

- Understanding ethical ideas and issues—such as justice, equity, respect, fairness, freedom and empathy;
- Considering the reasons behind actions and decisions;
- Exploring values, right and responsibilities from different points of view.

Some Scientists feel that the practice of Science does not need Ethics, as they say that Ethics cannot be taught to anyone, but ethical conduct is usually learnt through example, practice and osmosis. These notions pose hindrance in the serious study of the discipline 'Ethics' and its integration in Science (Resnik, 1998). Science is often treated as objective and value-free by many Scientists who condone the controversial and vexing ethical issues associated with various technologies. As a result, the Science curriculum which in a way translates the objectives and ideals set forth by the policymakers and other stakeholders also tends to obliterate such issues of ethical significance which goes against their rhetoric of so-called 'development'. Another argument that goes against introducing Ethics at higher education level is a claim put forward by some scientists that ethics could only be taught during

[2]Source: www.australiancurriculum.edu.au.

childhood, as that is the time when a child learns maximum and retains it throughout life. However, evidence from developmental psychology indicates that people continue to learn about ethics and moral reasoning throughout life (Rest, 1986). In fact, for building an ethical understanding the learners should continually engage and reflect upon instances, experiences and ethical dilemmas from their daily life so that they can be trained in the process of decision-making which is the ultimate goal.

The role of ethics in Science teaching is very well established by now, owing to the inclusion of ethical decision-making skills within the quality benchmarks of graduates in many countries of the world (QAA, 2002). However, the challenge is to determine who should teach it, how it should be taught and what curriculum should be included? Kohlberg (1958) had given a stage-wise theory of Moral Development in line with Piaget's Theory of Cognitive Development (1936) for describing the development of moral reasoning amongst individuals. The three different stages of moral development according to Kohlberg are as follows (refer to Fig. 11.1):

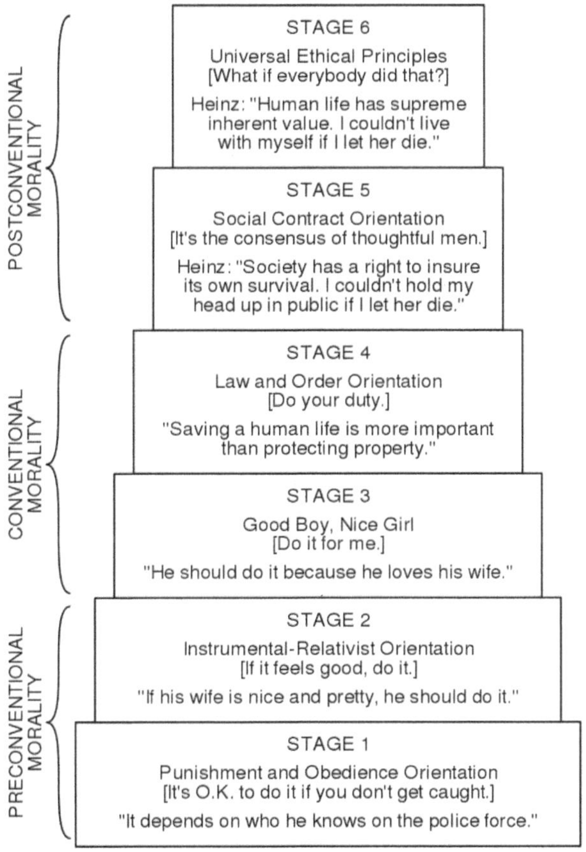

Fig. 11.1 Kohlberg's theory of moral development (*Source* https://en.wikipedia.org/wiki/Lawrence_Kohlberg's_stages_of_moral_development#/media/File:Kohlberg_Model_of_Moral_Development.svg (Accessed on 02.06.2019))

I. *Pre-conventional Stage*: This stage is characteristic of children and some adults too. Here, the morality of any action is determined by its consequences on their own self. This includes obedience- and punishment-related attributes. The stage is mostly driven by the self-interest of the individual.
II. *Conventional Stage*: This kind of moral reasoning is characteristic of adolescents and adults. Here, society's views and expectations gain predominance in deciding upon the morality of a particular action. Adherence to rules, laws and conventions is central to this stage.
III. *Post-conventional Stage*: This stage is characterized by the development of one's own ethical subjectivities and moral reasoning which may not be aligned to society's rules or standards. However, it has been stated that only a few people reach this level of abstract moral reasoning, and even for those who reach the stage, it is difficult to ascertain if they are consistent in it.

Thus, from the Moral Development theory (refer to Fig. 11.1), it becomes clear that individuals could be at very different levels of moral or ethical understanding. This is also a testimony to the fact that teaching ethics or morals through a didactic approach would not do much of help as would participatory and problem-based approaches.

11.3 Need for Assessing Ethical Understanding

The literature is replete with studies that disqualify 'Ethics' from the curriculum especially Science curriculum based on a simple reason that Ethics is a branch of Philosophy and has a different method of inquiry which is antagonistic to Scientific method. Also, there is a belief amongst the scientific community that ethics cannot be taught by professionals of Science, and if at all they do, it would be a disservice to the disciplinary rigour. Therefore, some studies have also emphasized upon the role of team or collaborative teaching, where both scientists and Philosophers/ Bioethicists collaborate on teaching a particular ethical issue related to Biosciences (Kabasenche, 2014). However, there has been a recent spurt in researches that claim that ethics is an integral part of Science and its processes and also that Science without Ethics cannot lead to humanitarian growth and development. As an offshoot of this, Universities have included Ethical Understanding in their Graduate Attributes (GA) and have also initiated courses in Bioethics for training the professionals in Science. Such courses offer 'Translation' of disciplinary understanding of real-life situations and ethical dilemmas (Barrie, 2006). The learning outcomes for ethical deliberation can be summarized as follows (Matchett, 2008, p. 32):

1. *Knowledge of*:
   A. An array of values, principles and ideals;
   B. Potential conflicts amongst those values, principles and ideals;

C. Facts that are especially relevant to ethical decisions in specific areas;
D. A core set of useful conceptual tools (for example, well-established ethical theories) and the reasons for their selection.

2. *Skills in*:

A. Multiple perspectives taking;
B. Formulating arguments that are logical, careful and clear;
C. Employing the tools identified in 1.D;
D. Applying standards that are commonly expected in both ordinary and professional social roles;
E. Analysing, evaluating or otherwise relating to any number of other tools and standards to those identified in 2.C and 1.D.

3. *Motivation and conviction*: that is, the conscious affirmation of and pattern of living habitually in accord with (one's) moral or ethical judgements.
4. *Implementation*: that is, the practical and emotional ability to carry out the course of action that (one) has judged ought to be done and is motivated to do.

The above pointers for ethical deliberation help in ascertaining and systematizing the integration of ethics within the curriculum and also help in setting criteria for assessment. The core value component which is placed at the top (1A) actually serves as a nucleus for achieving rest of the objectives. The above-laid criteria demarcate an area of ethical understanding and foresee its ramifications into the mainstream curriculum at both school education and higher education levels as they are very much embedded in the vision and mission statements of these institutions of learning. The direct implication would be better quality standards in education, development of holistic personality of learners along with character and virtues, resulting in peace and harmony that are much needed in today's day and age. Education in Morals and Ethics is a lifelong process and gets refined with age and experience.

11.4 Pedagogy of Assessing Ethical Understanding

As it is evident from the above discussion, assessment of ethical understanding is a difficult terrain and more implicit. Hence, it is important to envision a definitive way which would lead us towards a better understanding of the same. Pedagogy and assessment are intimately connected and rather than considering assessment as an end-term activity; it has to be more continuous and facilitate student learning in the process (refer to Fig. 11.2). Assessment for Learning (AfL) holds greater significance in learning as compared to Assessment of Learning (AoL) (Smith, 2016).

AfL can be effective only when tasks or activities designed are able to diagnose students' learning levels and their difficult areas and use of this knowledge by the teacher who would alter the strategies and plan interventions for improving learning

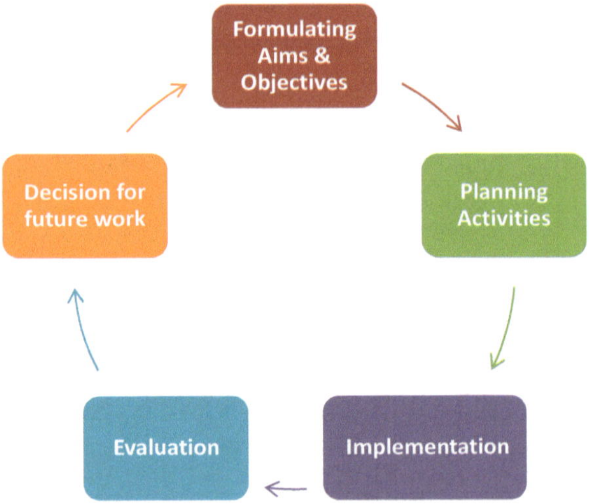

Fig. 11.2 Formative assessment cycle (*Source* Author)

(Black, 2016). AfL is also a practice for improving teacher's understanding of their learners and their ways of learning.

The domain of ethical understanding is situated within both cognitive and affective domains. There could be different ways of assessment of ethical understanding depending on the criteria and objectives set for the course, for example, a technique as simple as a 'reflective essay' could help in ascertaining the value basis and ethical stance of the students with respect to a particular ethical issue/problem. Balogh (2002) has suggested different alternative approaches for generating evidence for ethical understanding, such as case studies, recording of observations in media diaries, journals/logbooks, role plays, skits and debates. Each of these can be explained as follows.

11.4.1 Case Study

Students can be given certain case studies of ethical misconduct or the one requiring decision-making on the part of the learners based on bioethical issues or any of the Science processes. Students can analyse, reflect and write about the responses for these case studies either in groups or individually. Here, they need to apply their knowledge of some of the ethical theories and universal ethical principles in order to reach any conclusion. The task of the teacher is to evaluate the course of students' argumentation and discussion on the ethical issues that they have identified and also to keep a tab on the personal values and ethical understanding that the students display while reaching particular stance or conclusion. For instance, the case study given in Boxes 11.1 and 11.2 can be used to generate discussion and argumentation in the classroom and the students can be simultaneously assessed based on the quality of their arguments (refer to Appendix H).

Box 11.1 Case Study on Scientific Misconduct
The Baltimoore Affair

During the summer of 1991, the New York Times reported a scandal in scientific research conducted under the supervision of Nobel Prize winning scientist David Baltimore. The paper appeared in the 25 April 1986 issue of the journal 'Cell' and had listed the names of six authors. Baltimore supervised the research that was conducted in the lab; however, he did not carry out the experiments himself. The paper claimed that insertion of a foreign gene in a mouse could induce the production of antibodies from mouse's genes which are identical to the gene that was inserted. If this were true, then there is a possibility to control the immune system by inserting foreign genes which is a breakthrough research in the field of Medicine. However, a post-doctoral researcher working in the same institution, Margot O'Toole shared some contradictory findings and blew the whistle on this research by informing the review boards at MIT that led to further investigations of the research. The fraud was revealed by the investigations carried out by NIH although the researchers involved in the experiment maintained their innocence and unwariness about the whole issue. But, as a result of this, O'Toole found difficulty in getting further posting and work, as she became labelled as a 'troublemaker'. Baltimore, although holding a supervisory role, still resigned his Presidency from Rockefeller University in December 1992. He even asserted that many of the discrepancies in the results were less due to any manipulation and more due to sloppiness.

The Baltimore episode does raise many ethical concerns regarding the process and practice of Science, such as,

- Should Baltimore have paid closer attention to the research that was done under his supervision?
- Should his name appear as one of the authors of the paper if he himself did not conduct any experiment?
- Should Margot O'Toole have been given greater protection for her whistle blowing?
- Should poor recording be considered as an excuse for reporting false data and claims?

Box 11.2 Case Study on Cloning
Dolly—The Clone

(Source: https://www.flickr.com/photos/bettyvidal/6681950003)

On 23 February 1997, the Scottish scientists announced that they had cloned a sheep named 'Dolly' from the adult cells. This was for the first time that a mammal was produced out of adult cells without the natural process of mating and fertilization. Ian Wilmut, an Embryologist at the Roslin Institute, Edinburgh along with his colleagues cultivated cells from Ewe's udder in the laboratory, removed the nuclei from those cells, and used electric current to fuse these nuclei with the enucleated egg cell of the sheep. They then implanted the cells in Ewes' uteruses. Only 19 out of 277 embryos created in this fashion were viable and only one out of these was born (Wilmut, Schnieke, McWhir, Kind, & Campbell, 1997). This was a breakthrough research in the field of Biotechnology and according to some scientists could prove to be an antidote to many uncured diseases. Cloning technique when combined with gene therapy could have many useful applications in the fields such as Agriculture, Pharmaceuticals and Biotechnology. Wilmut had conducted this research in order to turn sheep into drug factories. The further applications as deemed by some scientists could be to make low-fat chickens, pigs that can serve as organ donors, super lactating cows and animals that produce human hormones (Resnik, 1998). However, this kind of scientific research raises many ethical concerns such as

- Is it fair to convert animals into factories just for the sake of human consumption?
- Animals also have their right to live and own will. Is it fair to manipulate and put them for human use?
- Reproduction is a natural process of production of new offspring. Is it ethical to interfere with the natural process and create offspring artificially without even contemplating the consequences?
- Dolly suffered a lot of pain before she was euthanized. Is it ethical to inflict so much pain and suffering to animals just for the sake of an experiment driven by human interests?
- This experiment could further even lead to human cloning. Does the cloning of people threaten the dignity, uniqueness and sacredness of human life?

11.4.2 Recording Observations in Media Diary

Students can be asked to maintain a media diary (refer to Table 11.1) wherein they will record their exposure to a particular media and the duration for which they use the particular platform. Teachers can share some resources related to ethical issues (Newspaper/magazine/articles/documentaries/short movies/advertisements/snippets). Students can keep a log of ethically problematic situation and, of course, their opinions and reflections on the same. This can be done as a semester-long exercise, and towards the end, each of the students can make a brief presentation about the lessons learnt from the exercise.

Table 11.1 Sample of media diary for assessing ethical understanding

Day	Time	Medium	Content	Place	Reflections/ role of bioethicist
Thursday	3.00 p.m.	Television	Movie showing gender bias	At home	
Friday	4.00 p.m.	YouTube	Video showing the farmer suicides due to GM crops	In car	
Saturday	11.00 a.m.	Internet	Facebook or Twitter posts related to animal experimentation	At home	
Saturday	5.00 p.m.	Newspaper	News columns on rising pollution levels due to human activity	In metro	
Sunday	12.00 p.m.	Magazine	Articles/advertisements on latest gadgets and technology	At home	
Sunday	3.00 p.m.	DVD	A movie, such as 'Brave New World', 'The Boys from Brazil' and 'Jurassic Park'	On laptop	
Monday	7.00 p.m.	Shared video	About the use or misuse of a particular technology	At home	
Tuesday	8.00 p.m.	YouTube	Documentaries on human-driven environmental disasters	In bus	
Wednesday	9.00 p.m.	Internet	Facebook post/blogging on current environmental issues	At home	

11.4.3 Maintaining a Journal

Maintaining a reflective journal by the students to record their day-to-day events is especially those involving contrasting values and decision-making by weighing the pros and cons of each situational outcome. This kind of journal writing involves 'reflection on action' which occurs after a particular action has taken place and the individual reflects over it in order to learn from the experience. 'Reflection on action' often feeds into 'reflection in action' which involves reflecting on the action while practicing it. Another term has also been given, such as 'knowing in action' that can come only from practice in reflective and reflexive thinking (Schon, 1983). Research has shown that students who engage in such kind of reflective thought processes acquire a centre stage in their own learning (Davies, 1995). Recording and describing a past event brings out the emotions and feelings associated with it and it recreates the entire episode that provides opportunities for its further exploration (Holmes, 1997). A student who is going through difficult times could

Table 11.2 Expectations from the teachers and students for journal writing

1.	Purpose of the journal	• To promote critical thinking, reflective thinking, problem-solving and divergent thinking amongst the learners • To develop their affective thinking and awareness of the ethical issues • To enable them to take a stance with respect to a bioethical issue and take action accordingly
2.	Expected format of the journal	• It could be hand-written/typed/designed using different media/software • Full sentences are required and issues to be explained elaboratively using examples, anecdotes, personal experiences, media reports, etc.
3.	Topics for writing journals	• Assigned topics related to Bioethical issues/recent reports/article/commentaries • Students' daily experiences • Students' chosen topics
4.	Extent of writing required	Depending on the issue taken up and individual writing skills of the learners but has to incorporate varied perspectives and arguments
5.	Assessment and evaluation of journals	• Journals can be checked by the teacher weekly/biweekly/monthly • Feedback can be given by writing some statements on the journal itself or can call a meeting to discuss one-to-one with the teacher • Grading can be done based on the quality of the journal

Adapted from Walker (2006)

record the experience of taking a tough decision and then reflecting over the reasons for choosing one particular action over another. Journals can also include some daily news items, case studies or recent media reports on a particular Bioethical issue and then a reflective entry about what the issue is about, its probable causes, roles of different stakeholders, impact on society, environment and oneself and whether it is right to take that research in the public domain.

> The role of the teacher could be to read the students' journals and provide them feedback and ask certain key questions related to their stance and decision with respect to a particular ethical issue (refer to Table 11.2 for Expectations from the Teacher).

11.4.4 Short Assignments and Learning Log

Students can be asked for writing short assignments, such as free-writing assignment at the beginning of the class which can help in triggering a discussion or as an end of the class activity where they can write a paragraph or two about their experience of learning about a new concept or issue.

Learning logs are tools to assess students' understanding in a stage-wise manner. The learning logs can be created either by the teachers or by the learners themselves, depending upon the level of the learners. The main aim of the learning log is to facilitate student learning of the concepts from simple to complex without any formal assessment or marks. Such a practice also allows the student to become equal partners in their own assessment so that they become aware of their own difficulties and weak points and can improve them in further tasks. Learning logs help promote meta-cognition amongst the learners so that they are able to better understand their own learning styles and strive to improve their learning. A learning log can be defined as

> The definition of a log as the daily record, the full record, a record of performance and a report on the construction of something (p. 41, Perritt Lee, 1997)

Logs can be used to introduce, develop, apply and evaluate classroom instruction and student learning. For the students also, logs can help in keeping a personal record of their learning, their reflection over the meaning-making and their extension and elaboration of the concepts (Perritt Lee, 1997). This could serve as a meaningful way of self-assessment by the students or peer-assessment while discussing the elements in their logs. These logs could also be assessed by the teachers by giving some written responses over them. Studies have shown that teachers' responses and detailed comments on the student logs encourage them to even write better (McIntosh & Draper, 2001).

The structure of a basic learning log can be taken as a two-column page (refer to Fig. 11.3) where on one side the students write down about the major concepts/ideas covered in the class or their questions/doubts/arguments, and on the other side, they can write the responses or explanation about those concepts/questions. This kind of an exercise would help the students in rationalizing their thought process and develop meta-cognitive awareness of their ways of learning and their difficulties and problems.

This kind of a learning log can very well be used for tracking ethical understanding amongst students wherein the students can be asked to pen down their ideas and responses to a particular ethical issue and the teacher can collect student learning logs on a weekly basis to assess their level of ethical understanding.

11.4.5 Cognitive Maps for Evaluating Students' Ethics Knowledge

Cognitive maps and concept maps are considered to be useful mapping techniques to demonstrate students' understanding of the concepts. These include mind maps (Budd, 2004), cognitive maps (Eden, 2004; Pinch, Sunley, & Macmillen, 2010), knowledge maps (McCagg & Dansereau, 1991) and concept maps (Novak, 1990, 1995). Concept maps can be of different types, depending upon the arrangement of concepts and their inter-relationship to each other, e.g. spider maps, having a central unifying factor and the related sub-themes radiating from it; hierarchy maps, where the most important or bigger concept is placed at the top followed by sub-concepts;

Name:________________ Date:________________

Learning Log: Basic Learning Log

TOPIC/BIG IDEA:

What I Learned (Concepts/Main Ideas/Key Details/Facts)	My Reaction (Your Thoughts/Questions/Arguments/Connections)

Fig. 11.3 A sample learning log (*Source* https://goalbookapp.com/toolkit/strategy/learning-logs)

flowchart, which depicts a process or steps; pictorial maps, etc. (All, Huycke, & Fisher, 2003). The assessment criteria for cognitive maps/concept maps include the following:

- Organization of the concepts presented;
- Considering competing ethical positions and multiple perspectives;
- Qualitative assessment of map morphology, including spokes, chains and nets (Kinchin & Hay, 2000);
- A high-content score reflects a deep understanding;
- Visuals added along with the content depicting better connection and clarity.

The concept maps can be hand-drawn as well as by using software available for creating concept maps on the computer (e.g. MindMaple). The students can make their maps as creative as they can think (refer to Figs. 11.4 and 11.5) by drawing inter-connections, relationships, adding propositions, examples, images, etc. In case of a concept map based on Bioethical issues, students can include different

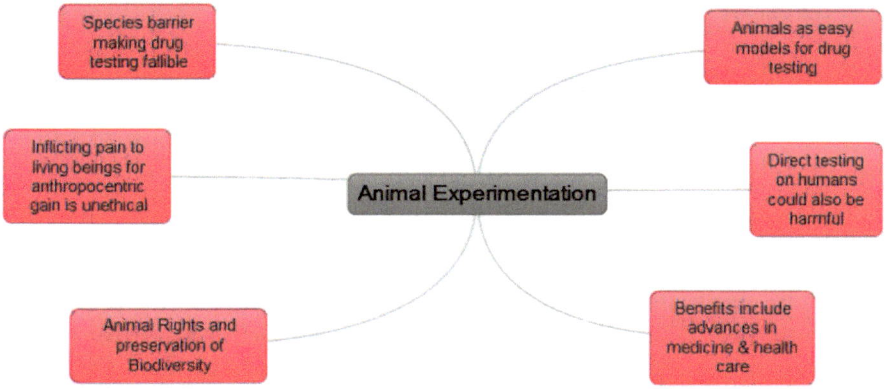

Fig. 11.4 Sample concept map on 'Animal Experimentation' (*Source* Author)

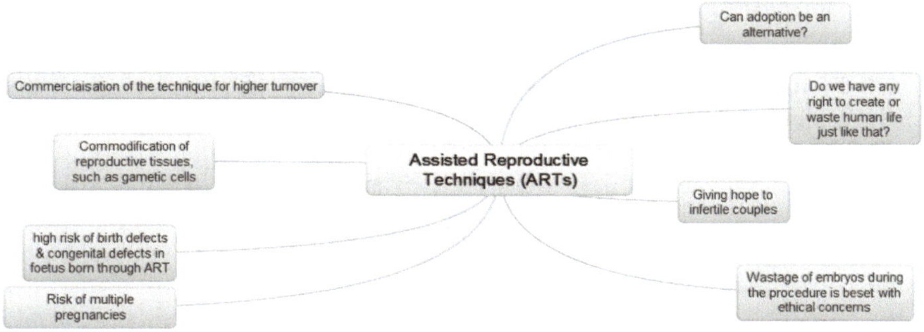

Fig. 11.5 Sample concept map on 'Assisted Reproductive Techniques' (*Source* Author)

arguments (pro-technology and anti-technology) and can easily weigh which side is better defended. This could also help in decision-making and reaching a stance with respect to a particular technology.

11.4.6 Blogging as a Way of Assessment of Students' Ethical Understanding

Blogging and creation of blogs can be a way to foster students' creativity and imagination. Blogging can be regarded as a technique for e-learning and blended learning by embedding ICT within the teaching–learning processes. Blogs can be created individually or in groups of desired number of students. The blogs can be created on Google Blogger, WordPress or any other online platform that offers the service. Studies have found that students learn better and reflect while they learn when they work collaboratively using e-learning tools (such as blogging) (Jackling,

Natoli, Siddique, & Sciulli, 2015). Such an e-learning technique can create a more learner-centred learning environment and address the needs of a heterogeneous group (O'Neill, Sing, & O'Donoghue, 2004). The use of learning through blogs promotes a constructivist approach to learning (Vygotsky, 1978) by building a discursive and interactive platform where students share their opinions and ideas and develop higher order thinking skills (Cameron, 2012). Students can be asked to develop blogs on the ethical concerns associated with different Bioethical issues wherein they can add and pool in evidence from different sources to provide more credence and validity to their work. They can be asked to post at least two to three blog entries per week and the teacher can assess them on a weekly basis and can grade them based on the quality of work and students' understanding about the ethical issues. This can also be done in a semester-long exercise mode where the students and teachers both can observe the gradation of their own thoughts and maturity in their ethical reasoning.

11.5 Conclusion

It has been well established that there are multiple ways to assess the ethical understanding of the students which is not only one of the graduate learning goals of a University or School but also a lifelong practice. Although, from the above discussion, it is evident that unlike any other topic or area of study, Bioethical issues and their understanding have to adopt a non-conventional route for assessment. Research has pointed out towards the conspicuous neglect of ethical issues within the curriculum, including their teaching–learning and assessment at all levels of education. There have been specialized courses designed specifically to address these Bioethical issues but only in a few selected universities and programmes. The present chapter delineated the importance, rationale, objectives and strategies for assessment of ethical understanding amongst students. The different strategies and techniques, such as case studies, media diaries, journal writing, learning log and concept/mind mapping are reflective in nature and promote more self-assessment rather than teacher-led assessment. These are helpful learning strategies that enable young learners to take charge of their own learning and improve it each day. Although the role of teacher cannot be negated here, the teacher is the one who is responsible for guiding the learners on the different strategies and solving their doubts or hindrances related to the technique. It provides formative feedback to the teacher as well, as she is able to track the understanding of her learners and plan her teaching accordingly.

References

All, A. C., Huycke, L. I., & Fisher, M. J. (2003). Instructional tools for nursing education: Concept maps. *Nursing Education Perspectives, 24,* 311–317.

Balogh, D. W. (2002). *Teaching tips-teaching ethics across the psychology curriculum.* Retrieved from Association for Psychological Science: https://www.psychologicalscience.org/teaching/tips/tips_0902.cfm.

Barrie, S. (2006). Understanding what we mean by the generic attributes of graduates. *Higher Education, 51*(2), 215–241.
Black, P. (2016). The role of assessment in pedagogy—And why validity matters. In D. Wyse, L. Hayward, & J. Pandya (Eds.), *The Sage handbook of curriculum, pedagogy and assessment* (pp. 725–739). London: Sage Reference.
Budd, J. W. (2004). Mind maps as classroom exercises. *Journal of Economic Education*, 35–46.
Cameron, M. (2012). 'Economics with training wheels': Using blogs in teaching and assessing introductory economics. *The Journal of Economic Education, 43*(4), 397–407. https://doi.org/10.1080/00220485.2012.714316.
Davies, E. (1995). Reflective practice: A focus for caring. *Journal of Nursing Education, 34,* 167–174.
Eden, C. (2004). Analyzing cognitive maps to help structure issues or problems. *European Journal of Operational Research, 159,* 673–686.
Holmes, V. (1997). Grading journals in clinical practice: A delicate issue. *Journal of Nursing Education, 36,* 489–492.
Jackling, B., Natoli, R., Siddique, S., & Sciulli, N. (2015). Student attitudes to blogs: A case study of reflective and collaborative learning. *Assessment & Evaluation in Higher Education, 40*(4), 542–556. https://doi.org/10.1080/02602938.2014.931926.
Jones, M., van Kessel, G., Swisher, L., Beckstead, J., & Edwards, I. (2014). Cognitive maps and the structure of observed learning outcome assessment of physiotherapy students' ethical reasoning knowledge. *Assessment & Evaluation in Higher Education, 39*(1), 1–20. https://doi.org/10.1080/02602938.2013.772951.
Kabasenche, W. P. (2014). [The ethics of] Teaching science and ethics: A collaborative proposal. *Journal of Microbiology and Biology Education*, 135–138.
Kinchin, I. M., & Hay, D. B. (2000). How a qualitative approach to concept map analysis can be used to aid learning by illustrating patterns of conceptual development. *Educational Research, 42,* 43–57.
Kohlberg, L. (1958). *The development of modes of moral thinking and choice in the years 10 to16 (1st ed.).* Ph.D. Thesis, University of Chicago.
Matchett, N. J. (2008). Ethics across the curriculum. In S. L. Moore (Ed.), *New directions for higher education—Practical approaches to ethics for colleges and universities* (pp. 25–38). San Francisco: Jossey-Bass.
McCagg, E., & Dansereau, D. (1991). A convergent paradigm for examining knowledge mapping as a learning strategy. *Journal of Educational Research, 84,* 317–324.
McIntosh, M., & Draper, R. J. (2001). Using learning logs in mathematics—Writing to learn. *Mathematics Teacher, 94,* 554–557.
Novak, J. D. (1990). Concept maps and Vee diagrams: Two metacognitive tools to facilitate meaningful learning. *Instructional Science, 19,* 29–52.
Novak, J. D. (1995). Concept mapping to facilitate teaching and learning. *Prospects: Quarterly Review of Comparative Education, 93*, 79–86.
O'Neill, K., Sing, G., & O'Donoghue, J. (2004). Implementing e-learning programmes for higher education: A review of the literature. *Journal of Information Technology Education, 3,* 313–323.
Perritt Lee, E. (1997). The learning response log—An assessment tool. *The English Journal, 86*(1), 41–44.
Piaget, J. (1936). *Origins of intelligence in the child.* London: Routledge & Kegan Paul.
Pinch, S., Sunley, P., & Macmillen, J. (2010). Cognitive mapping of creative practice: A case study of three English design agencies. *Geoforum, 41,* 377–387.
QAA. (2002). *Benchmarking academic standards: The biosciences*. Quality Assurance Agency.
Resnik, D. B. (1998). *The ethics of science—An introduction*. New York: Routledge.
Rest, J. (1986). *Moral development*. New York: Praeger.
Schon, D. (1983). *The reflective practitioner*. London: Temple Smith.
Smith, K. (2016). Assessment for learning: A pedagogical tool. In W. Dominic, L. Hayward, & J. Pandya (Eds.), *The Sage handbook of curriculum, pedagogy and assessment* (pp. 740–755). London: Sage reference.

Swisher, L. L. (2007). The problem of measurement in ethics education. In *Symposium: Educating for Moral Action: Creating a Global Dialogue*. World Confederation for Physical Therapy, Vancouver, Canada.
Tyler, R. W. (1949). *Basic principles of curriculum and instruction*. Chicago, IL: University of Chicago Press.
Vygotsky, L. S. (1978). In M. Cole (Ed.), *Mind in society: The development of higher psychological processes*. Cambridge, MA: Harvard University Press.
Walker, S. E. (2006). Journal writing as a teaching technique to promote reflection. *Journal of Athletic Training, 41*(2), 216–221. Retrieved from https://www.ncbi.nlm.nih.gov/pmc/articles/PMC1472640/pdf/i1062-6050-41-2-216.pdf.
Wilmut, I., Schnieke, A., McWhir, J., Kind, A., & Campbell, K. (1997). Viable offspring derived from fetal and adult mammalian cells [letter]. *Nature*, 769–771.

Websites

https://ethicstool.com/.

Appendix A
Categories of Ethical Issues in Biological Sciences and Applied Biotechnology

Category no.	Category name/subcategories	Major ethical issues
1.	**Animal experimentation**	1.1 Ethics and care of dealing with live animals
		1.2 Ethics and care of keeping animals in the lab
		1.3 Ethics of handling animals while doing experiments
		1.4 Minimum wastage and care for animal life
		1.5 Observance of animal rights
2.	**Sexual reproduction**	
	2.1 Sex-determination via amniocentesis	2.1.1 Principle of equality
		2.1.2 Sensitivity towards life and care for it. (right to life)
		2.1.3 Inherent biases in the society (patriarchy)
		2.1.4 Medical profession ethics
	2.2 Use of contraceptives	2.2.1 Interference with the nature's rules
		2.2.2 Clash between the scientific argument and church's position
		2.2.3 Health implications
		2.2.4 Contraceptives as abortifacient
		2.2.5 Promiscuity and impact on relationships and family bonding
	2.3 MTP or abortion	2.3.1 Right to birth and right to life
		2.3.2 Foetus as a potential adult
		2.3.3 Principle of sentience
		2.3.4 Instances where MTP is ethically appropriate decision

(continued)

A. Saxena, *Ethics in Science*, https://doi.org/10.1007/978-981-32-9009-9

(continued)

Category no.	Category name/subcategories	Major ethical issues
	2.4 IVF and other assisted reproductive techniques	2.4.1 Evading stigmatization
		2.4.2 Wastage of embryos during IVF
		2.4.3 Use of IVF even by the fertile
		2.4.4 Principle of utilitarianism
		2.4.5 Increasing divide between haves and have nots
3.	**Biotechnology**	
	3.1 GM crops and associated ethical issues	3.1.1 Unpredictability of the outcomes
		3.1.2 Maintenance of safety standards
		3.1.3 Threat to farmer's autonomy
		3.1.4 Gene patents and biopiracy
		3.1.5 Threat to biodiversity
		3.1.6 Emergence of superweeds and superbugs
		3.1.7 GM crops versus organic farming
	3.2 Genetically Modified Organisms (GMOs) and associated ethical issues	3.2.1 Misuse of GMOs as bioweapons
		3.2.2 Limitation of principle of utilitarianism
		3.2.3 Case-by-case analysis as the best ethical approach
		3.2.4 Pain and suffering experienced by the concerned animals
		3.2.5 'Playing God' and tinkering with the blueprints of life forms
		3.2.6 Intrinsic respect and consideration for other life forms
		3.2.7 'Transgenesis' leading towards breaching of species barrier
	3.3 Human genomic project and associated ethical issues	3.3.1 Principle of privacy and confidentiality of genetic information
		3.3.2 Playing with nature's diversity and gene pool
		3.3.3 Issue of risking human life
		3.3.4 Issue of informed consent
		3.3.5 Issue of patenting genetic information
	3.4 Stem-cell therapy and associated ethical issues	3.4.1 Issue of potentiality
		3.4.2 Issue of human rights

(continued)

(continued)

Category no.	Category name/subcategories	Major ethical issues
4.	**Environmental ecology**	
	4.1 Use and overuse of natural resources	4.1.1 Overconsumption of natural resources
		4.1.2 Responsibility towards the future generations
		4.1.3 Principle of equality and equal distribution of resources
	4.2 Conservation of biodiversity	4.2.1 Anthropocentric view of deriving maximum benefit
		4.2.2 Biocentric and cosmocentric view
		4.2.3 Principle of deep ecology and shallow ecology
	4.3 Pollution and global warming	4.3.1 Disturbing atmospheric balance
		4.3.2 Disturbing the growth of flora and fauna
		4.3.3 Conflict between the developed and developing countries regarding their carbon footprints
		4.3.4 Dilemma between lesser polluting technology or a more stabilizing one

Appendix B
GRID for Topic-Wise Analysis of Ethical Issues in the K-12 Biology Textbook

Type of ethical issue	The ethical issues can be of various types such as medical/socio-ethical/politico-legal/emotional/etc.
Category no.	This is the code number being assigned to each kind of ethical issue in the list of categories prepared prior to analysis of the textbook. This will provide a kind of tally as to how many categories have been included in the textbook and how many are left out
Broad area	This refers to the main branch of Biological Sciences to which a particular ethical issue belongs
Length of issue covered in textbook	This is the amount of space given to the explanation of a particular ethical issue in the textbook, which can be determined in terms of words, sentences, paragraphs or pages
Reference to any ethical theory	The ethical theories being referred to here include the following: **Virtue Ethical Theory**: This particular theory stems largely from the philosophy of Plato and lays down the norms for good or bad behaviour in terms of identification and practicing of certain values and ideals, thus making it normative in character **Utilitarian Theory**: This theory proposes that happiness lies in observance of the 'principle of beneficence', which entails maximum benefit to maximum people **Consequentialist Theory**: This constitutes a part of the Utilitarian theory and can be separately viewed in terms of the end product/result of any action that holds critical importance than the action in itself. Hence, the theory stresses the role of ends rather than the means **Deontological Theory**: This theory derives its significance from Immanuel Kant's work on 'Categorical Imperative', and hence regards actions as duty-bound rather than ends driven
Arguments raised	The situation of dilemma arises where different groups of people hold diverse opinions regarding a particular issue. These arguments can be from the realm of social, political, scientific, religious or cultural arena. Taking into account all of these arguments is important so as to generate an informed understanding of a particular issue and also for a balanced decision-making on the same

(continued)

A. Saxena, *Ethics in Science*, https://doi.org/10.1007/978-981-32-9009-9

(continued)

Type of ethical issue	The ethical issues can be of various types such as medical/socio-ethical/politico-legal/emotional/etc.
Kind of understanding fostered	Learner Understanding or Comprehension is the outcome of any text. Our present focus is on generating an ethical understanding amongst the students by way of textual content, ability to weigh the pros and cons of a particular issue by taking account of all the arguments and a sound decision-making on these controversial issues. The four different levels of comprehension can be proposed depending on the extent of impact of textual content on the minds of young learners: • Factual • Conceptual • Ethical dilemma • Ethical decision-making

Appendix C
Proforma for Survey

Name:

Mobile Number and E-mail ID:

Qualifications (along with affiliated institutes if possible):

Designation:

Experience in Teaching: (Senior Secondary): (Undergraduate level):

School/College:

Subjects offered for Teaching:

Areas of Interest in Teaching:

Hobbies:

References used for teaching: (textbook/any private publication books/e-books/newspaper articles/magazines/journals/Google search/Wikipedia)

Any other Reference: (Please specify):

Most favoured teaching-learning approach:

- Lecture-mode
- Lecture + Discussion
- Interactive and more of student participation
- Using Audio-visual materials
- Inquiry approach (more of didactic and question-answer sessions)
- Student Presentations
- Any other (Please specify)

Any major reflection so far about the kind of teaching-learning approaches at higher secondary/undergrad level (do you see them as appropriate and unchangeable or do you want to make any modifications to the existing schemes and procedures of teaching-learning)?

In what way do you feel that students understand and learn better and why do you feel so (please give evidences from your own class-room experiences)?

Have you noticed any particular students' misconception (a false or erroneous explanation or reason to any incident or process) related to a specific topic? Please elaborate upon the specific topic and the related misconception that student had and also how you dealt with it.

What are your general perceptions about the kind of teaching-learning activities that go on in your school/college? (Express your satisfaction or dissatisfaction by giving evidences and examples).

A. Saxena, *Ethics in Science*, https://doi.org/10.1007/978-981-32-9009-9

Appendix D
Questionnaire for Teachers and University Professors

1. What are your major objectives of teaching Biology and how is it different from teaching of other subjects?
2. What do you understand by 'Scientific Literacy'?
3. Do you consider it as an important goal of science education, if yes, then how do you achieve it?
4. How far do you agree with the following statement: "Future citizens will need to be informed consumers of technology, science, sociology and ethics, to name a few?" (Gunn et al., 2008)
5. Is there any place for 'Ethics' in science? Please elaborate upon your perception by giving evidence.
6. Do you think ethical issues have been addressed in the curriculum of Biological Sciences at senior secondary/undergrad level? Please elaborate.
7. Why are these issues termed as 'ethical issues' and not as science and technology issues?
8. How do you tackle these issues in the classroom? (any specific teaching–learning approach adopted).

Reference

Gunn, T. M., Grigg, L. M. & Pomahac, G. (2008). Critical thinking in Science Education: Can Bioethical issues and questioning strategies increase scientific understandings? *Journal of Educational Thought (JET)*, *42*(2), 165–183.

A. Saxena, *Ethics in Science*, https://doi.org/10.1007/978-981-32-9009-9

Appendix E
A Tool to Analyse Classroom Processes in Dealing with Ethical Issues in Biological Sciences

A. Saxena, *Ethics in Science*, https://doi.org/10.1007/978-981-32-9009-9

S. No.	Scale						
1.	Introduction to the topic	Beginning with a summary of previous class	Beginning with a case/vignette/newspaper article related to the topic	Connecting with daily life incidents	Plainly introducing the topic (no background)	Any other	Remarks
2.	Kind of examples/analogies used	From daily life and personal experiences	From scientific researches quoted in newspapers/journals	From the textbook (if any)	From television and other media	Any other	
3.	Number of ethical arguments raised	None	1–2	3–4	5–6	7–8	
4.	Type of ethical arguments raised (along with some evidence)	**Virtue based** (evidence in classroom episode/interviews)	**Consequentialist** (evidence in classroom episode/interviews)	**Utilitarian** (evidence in classroom episode/interviews)	**Deontological** (evidence in classroom episode/interviews)	Any other (evidence in classroom episode/interviews)	
5.	Treatment given to ethical arguments	N.A	Just listed	Listed and discussed	Listed, discussed and debated	Listed, discussed, debated and resolved	
6.	Students' contribution to ethical arguments	No initiative	Just a few students active in discussion	Most of the students participating in the discussion	Students participating and contributing in discussion	Students try to quote evidence in favour/against any argument	
7.	Classroom time devoted to the teaching of ethical issues	No time devoted	<5 min	5–10 min	10–15 min	>15 min	
8.	Kind of resolution reached	Teacher maintains a neutral stand	Teacher takes a pro-technology stand and students abide by it	A dilemma created amongst students	Students left with a question	Students and teacher reach a consensus	Any other
9.	Major pedagogical approach adopted	Lecture	Discussion and debate	Using audio-visual aids/educomp as an add on without any discussion	Using audio-visual aids/educomp followed by discussion	Quoting vignettes or cases interspersed with discussion	Any other
10.	Kind of ethical understanding aimed at by the teachers	Factual understanding	Conceptual level Of understanding	Application level of understanding	Ethical dilemma level Of understanding	Ethical decision-making level of understanding	Any other

Appendix F
List of Codes Used in Coding Data

1. **Subject Matter Knowledge: (SMK)**: This represents the knowledge category of the case with respect to knowledge of the subject, content knowledge, knowledge about the subject's inter-relations with other allied subjects (Integrated subject), as well as applications of subject knowledge in daily life. The subject matter knowledge might not just include the knowledge given in the text, but the overall knowledge on a particular topic or area, that may include knowledge from latest newspaper articles, journals, magazines, TV programmes, etc.
2. **Teacher's Beliefs: (TB)**: Beliefs as a category of teacher's knowledge is very often confused with knowledge in general and specifically subject matter knowledge (SMK) here. However, the two have a slight difference, due to which these have been treated as different categories of knowledge. TB in the present study refers to teacher's conceptions, notions, views or maybe conventional beliefs about a particular ethical issue. This is a very dynamic category of teacher's knowledge in terms of interpersonal variations in the belief structures of the teachers with respect to a single ethical issue.
3. **Knowledge of Transactional Strategies: (KTS)**: This represents the category of teacher's knowledge about the different ways of transacting a topic in the classroom and includes the method of teaching–learning adopted, such as lecture method, lecture with student participation, completely interactive or inquiry based, discussion oriented, problem-solving, etc. This also takes into account teacher's use of examples, analogies, PowerPoint presentations, transparencies, documentaries, illustrations, use of media, etc.
4. **Knowledge of the Curriculum: (KCur)**: This includes teacher's knowledge about the curricular goals, short-term and long-term objectives, framing of the science curricula and the hidden agenda of sociopolitical structures in the society, place of ethical issues within the curriculum and the treatment given to them.

A. Saxena, *Ethics in Science*, https://doi.org/10.1007/978-981-32-9009-9

5. **Knowledge of Students' Understanding: (KSU)**: This refers to teacher's knowledge about the ways students understand better, their difficulties in understanding of particular topics, their misconceptions and common mistakes made by students within the topic boundary. This is usually reflected within teacher's ways of dealing with the topic, the method chosen and reason for the same, the tasks given to the students, the kind of examples used, etc.
6. **Knowledge of Context: (KCon)**: This represents teacher's knowledge about the institution's environment, administration, infrastructure, about classroom culture, knowledge about the broader community, its practices, that could in anyway hinder or facilitate teaching–learning.
7. **Knowledge of Assessment: (KA)**: This refers to the different assessment practices adopted by the teacher with respect to assessing student understanding vis-à-vis ethical issues in Biological Sciences. The assessment could take various forms such as student feedback, responses in classroom interactions, organizing debates and discussions, designing any specific activity for the same.
8. **Knowledge about role of ethics in science: (KET)**: This is the teacher's knowledge about the various ethical theories, such as virtue, utilitarian, deontological, etc. This might not exist in such complex terminologies but may be implicit in the statements made by them while taking a class, or during an interview session; hence, the general knowledge about ethics is also considered under this category.
9. **Orientation towards Science Teaching: (OST)**: This includes teacher's own motives and purposes of teaching science which could be very different from the general objectives of science teaching. This category of knowledge takes into account teacher's personal interests, motivation and criteria for teaching science.
10. **Knowledge of Ethical Argumentation (raising arguments/initiating argumentation/resolving): (KArg)**: This refers to teacher's knowledge about the valid ethical arguments with respect to a specific ethical issue. The arguments can be also in the form of a chain such that one argument initiated by the teacher breeds another argument from the students' side, thus initiating an ethical discourse in the classroom.
11. **Knowledge about Nature of Science: (KNOS)**: As is evident from the category name, it refers to the teacher's nature of science understanding which includes the various sub-themes, some of which have been covered in the present study.

Appendix G
Elements for Classroom Discourse Analysis

Teacher's knowledge	Nature of teacher's questions	Nature of students' responses	Focus of discourse	Teacher's role
Social context Including knowledge about outside social context and how it impacts classroom discussions	Open-ended (inviting diverse students' responses)	Students' Beliefs (epistemic self-portrait) Based on pre-conceived notions or strongly held beliefs based on one's culture and background (searching for metaphors, misconceptions and alternative conceptions)	Teacher-centred Teacher does all the talking in the class with minimal to negligible student participation	Role of teacher in a discussion Giving space for students' responses Silent intervals (to promote thinking and reflection)
Interactional context Frequency of turntaking between teacher and students, relationship between teacher and students, freedom of expression reflected in students' responses	Linguistically simplistic (language of questions simple and focussed)	Textual Based on knowledge about theories or the scientific process (which knowledge is valued by students and why the given theory is not challenged)	Teacher mediated Teacher leads the discussion in the classroom by asking questions, validating students' responses, directing the course of discussion in a predetermined direction	Direction of talk From a simplest idea to a complex one

(continued)

A. Saxena, *Ethics in Science*, https://doi.org/10.1007/978-981-32-9009-9

(continued)

Teacher's knowledge	Nature of teacher's questions	Nature of students' responses	Focus of discourse	Teacher's role
Personal agency Personal choices over content for the class, choice of methods, explanations, examples, analogies, etc.	Contextualized (teacher's questions well adapted to the context of learners and promote their engagement with the topic/issue)	Individual and social construction Responses build there and then based on collaborative thought, imagination and creativity (to look for students' process of theory building)	Student driven Students participate actively in classroom discussion focussing more on student–student interactions rather than teacher–student interaction	Component of subjectivity Portraying science and scientific knowledge as a value-laden practice involving subjectivity, researcher's background, biases, emotions, etc. Using may be, can be, etc. while explaining

Appendix H
Rubric for Assessing the Quality of Argumentation

Toulmin's model includes six components based on which any argument can be dissected and analysed (web.cn.edu/kwheeler/documents/Toulmin.pdf.)

1. *Data*: This represents the facts or evidence on which an argument is based.
2. *Claim*: This represents the statement or the argument that defines the proposed claim (or thesis).
3. *Warrants*: These are general hypothetical (often implicit) or logical statements that provide a bridge between the data and the claim, making the argument more strong.
4. *Qualifiers*: These act as limiting factors as they limit the validity or feasibility of a particular claim. It also proposes certain conditions under which the argument holds true.
5. *Rebuttals*: These refer to the statements or conditions under which the claim does not hold true such as providing some counterarguments for the proposed claim proving the conditionality of the claim.
6. *Backing*: This represents some theoretical justifications that support the warrant and not necessarily the main argument.

Levels of argument	Quality of argument
Level 1	Argument consisting of a simple claim based on a personal belief without any evidence/data/warrant/backing/rebuttal
Level 2	Argument consisting of a claim supported with data/evidence/backing which provides a valid reason for the claim but having no rebuttal
Level 3	Argument with a series of claims or counterclaims with data/warrant/backing that support the claim but having only a weakly identifiable rebuttal which indicate a pre-ponderance towards the validity of claim

(continued)

A. Saxena, *Ethics in Science*, https://doi.org/10.1007/978-981-32-9009-9

(continued)

Levels of argument	Quality of argument
Level 4	Argument having several claims and counterclaims supported with evidence and data and having a clear rebuttal which refutes the claim. Such an argument can also indicate a state of ethical dilemma
Level 5	An elaborated argument with more than one rebuttal indicating a true engagement with the argument and approximation to a consensus or truth (leading towards ethical decision-making)

Adapted from Erduran et al. (Erduran, S., Simon, S., & Osborne, J. (2004). TAPping into argumentation: Developments in the application of Toulmin's Argument Pattern for studying science discourse. *Science Education*, *88*, 915–933), p. 928

Glossary

Amniocentesis It is a process in which amniotic fluid is sampled using a hollow needle inserted into the uterus, to screen for abnormalities in the developing foetus

Autosomal Recessive Disorder It is a genetic disorder that is usually transmitted if both the parents are carriers of the mutated gene, then there are at least 25% chances that the offspring will be affected by the disorder, for example, Sickle Cell Anaemia, Cystic Fibrosis, etc.

Autosome It is any chromosome other than the sex chromosome

Bilirubin It is a yellow-coloured pigment that is formed in the normal catabolic pathway that breaks down heme in vertebrates. This catabolism is a necessary process in the body's clearance of waste products that arise from the destruction of aged or abnormal red blood cells

Biodiversity Biodiversity refers to the variety and variability of life on Earth. Biodiversity is typically a measure of variation at the genetic, species and ecosystem level

Biopiracy It is the practice of commercially exploiting naturally occurring biochemical or genetic material, especially by obtaining patents that restrict its future use, while failing to pay fair compensation to the community from which it originates

Bioprospecting It is the process of discovery and commercialization of new products based on biological resources. These resources or compounds can be important for and useful in many fields, including pharmaceuticals, agriculture, bioremediation and nanotechnology, among others

Chorionic Villus Sampling (CVS) It is a form of prenatal diagnosis to determine chromosomal or genetic disorders in the fetus. It entails sampling of the chorionic villus and testing it for chromosomal abnormalities

A. Saxena, *Ethics in Science*, https://doi.org/10.1007/978-981-32-9009-9

Chromosomal Disorder It is an abnormal condition due to something unusual in an individual's chromosomes

Chromosome It is the heredity-bearing gene carrier of the living cell, derived from chromatin, and largely consists of nucleoproteins, the nucleic acid being DNA. The unit of genetic information is the gene, and each chromosome may be regarded as composing a number of genes

Down's Syndrome It is a chromosome disorder caused by the presence of an extra copy of chromosome 21

Dysmenorrhea It is the medical term for menstrual cramps which are caused by uterine contractions

Endometrium It is the inner epithelial layer, along with its mucous membrane, of the mammalian uterus

Eugenics Eugenics is a science that aims to improve the genetic quality of a human population by excluding certain genetic groups judged to be inferior and promoting other genetic groups judged to be superior

Evolution Evolution is change in the heritable characteristics of biological populations over successive generations. These characteristics are the expressions of genes that are passed on from parent to offspring during reproduction

Gamete Also known as a germ cell is a reproduction cell, usually haploid and sexually differentiated. The female gamete (or ovum) unites with the male gamete (or spermatozoon) during fertilization to produce a zygote, which develops into a new individual

Gene Therapy It is the therapeutic delivery of nucleic acid into a patient's cell as a drug to treat a certain disease

Gene It is a hereditary factor or a unit of the material of inheritance. It is responsible for much of the similarity that occurs between organisms related by descent. It is a complex and diverse material that gives rise to accurate copies of itself, units of which are passed from parent to offspring. It is stable in its properties and profoundly affects every aspect of the individual organism containing it

Genetic Engineering It is the artificial alteration of genetic makeup of cells, customarily excluding selective breeding

Genetic Modification This is a process of altering or modifying the genetic makeup of an organism. This was done initially through indirect methods, such as by selective breeding of plants and animals, but with new technology it can now be done even directly

GIFT (Gamete Intra-Fallopian Tube Transfer) It involves the transfer of freshly recovered ova and conditioned spermatozoa into the fallopian tubes

Human Genome It represents all the genes of a human individual

ICSI (Intra-Cytoplasmic Sperm Injection) It is a technique wherein the spermatozoon is mechanically inserted into the oocyte

Implantation It is the stage of pregnancy at which the embryo adheres to the wall of the uterus. At this stage of prenatal development, the conceptus is called a blastocyst. It is by this adhesion that the embryo receives oxygen and nutrients from the mother to be able to grow

IVF or In Vitro Fertilization It is a technique wherein the fertilization takes place outside the uterus, i.e. in vitro, where the development of the embryo takes place for some time after which it is transferred into the uterus

Oliguria It is defined as a urine output that is less than normal, which can be 1 mL/kg/h in infants, less than 0.5 mL/kg/h in children and less than 400 mL or 500 mL per 24 h in adults—this equals 17 or 21 mL/hour

Ovary It is a glandular organ giving rise to ova (egg). In animals, ovary not only produces eggs but also releases sex hormones, such as Estrogen, Luteinizing hormone and Progesterone

Patenting A patent is a form of intellectual property that gives its owner the legal right to exclude others from making, using, selling and importing an invention for a limited period of years, in exchange for publishing an enabling public disclosure of the invention. The act of filing a patent to secure one's legal right over a particular invention or discovery is called as 'patenting'

Prothrombin Prothrombin is a glycoprotein (carbohydrate–protein compound) present in blood plasma and is an essential component of the blood-clotting mechanism

Salpingitis It refers to a medical condition that involves inflammation of the fallopian tubes due to bacterial infection. It is one of the causes of infertility among females as it can damage the fallopian tubes

Sex-Determination It refers to a system that determines the development of sexual characteristics in an organism. In humans, XX/XY sex-determination system is found, where XX chromosome pair results in the development of female fetus, while XY results in a male fetus

Sex-Linked Disorder Sex-linked diseases are passed down through families through one of the X or Y chromosomes. X and Y are sex chromosomes

SSI (Socio-Scientific) issues SSI refers to the controversial social issues that relate to Science and emphasises the role of both scientific and moral reasoning in promoting better understanding as well as in developing scientific literacy among masses

Stem Cells A stem cell is a cell having a unique ability to develop into any specialized cell types in the body

STS (Science, Technology and Society) issues STS issues are the issues that arise at the interface of science, technology and society. It is a study of how society, politics and culture affect scientific research and technological innovation, and how these in turn affect society, politics and culture

Surrogacy Surrogacy is an arrangement, often supported by a legal agreement, whereby a woman agrees to become pregnant and give birth to a child for another person who is or will become the parent of the child

ZIFT (Zygote Intra-Fallopian Tube Transfer) It is a technique wherein the zygote is directly transferred into the fallopian tubes of the uterus of the mother